Routle

T0231171

RELI
THE SCIENCES
OF LIFE

RELIGION AND THE SCIENCES OF LIFE

WITH OTHER ESSAYS ON ALLIED TOPICS

by
WILLIAM McDOUGALL, F.R.S.

First published in 1934 by Methuen & Co. Ltd.

This edition first published in 2018 by Routledge
2 Park Square, Milton Park, Abingdon, Oxon, OX14 4RN
and by Routledge
711 Third Avenue, New York, NY 10017

Routledge is an imprint of the Taylor & Francis Group, an informa business

Publisher's Note
The publisher has gone to great lengths to ensure the quality of this reprint but
points out that some imperfections in the original copies may be apparent.

Disclaimer
The publisher has made every effort to trace copyright holders and welcomes
correspondence from those they have been unable to contact.
A Library of Congress record exists under ISBN: 34041840

ISBN 13: 978-1-138-55698-0 (hbk)
ISBN 13: 978-1-138-56528-9 (pbk)
ISBN 13: 978-1-315-12289-2 (ebk)

RELIGION AND THE SCIENCES OF LIFE

by

WILLIAM McDOUGALL

METHUEN & CO. LTD. LONDON

RELIGION AND
THE SCIENCES OF LIFE

RELIGION AND THE SCIENCES OF LIFE

WITH OTHER ESSAYS ON ALLIED TOPICS

by

WILLIAM McDOUGALL, F.R.S.

METHUEN & CO. LTD.
36 Essex Street W.C.
LONDON

First Published in 1934

PRINTED IN GREAT BRITAIN

PREFACE

THIS volume collects a few of the many essays and addresses I have scattered in various journals and magazines during the last forty years, together with three that have not previously appeared in their present form. I am especially indebted to Messrs. Methuen & Co., and to Messrs. Kegan Paul & Co. for their kind permission to include 'Ethics of Nationalism' and 'World Chaos'; for these are very condensed presentations of the substance of volumes published by them. For permission to republish here I return thanks also to the editors and publishers of the following: *The South Atlantic Quarterly* ('Religion and the Sciences of Life'); *Philosophy* ('Mechanism, Purpose, and the New Freedom' and 'Apollonian and Dionysian Theories of Man'); *The Harvard Graduates' Magazine* ('The Need for Psychical Research'); *The Case For and Against Psychical Research* ('Psychical Research as a University Study'); *Scribner's Magazine* ('The Island of Eugenia'); *The Sociological Review* ('Family Allowances, 1906'); *Character and Personality* ('Family Allowances, 1933'); *The Forum* ('Was Darwin Wrong?'); *The Edinburgh Review* ('Our Neglect of Psychology').

In selecting these essays I have been guided partly by the desire to present matter likely to be of interest to the general reader; but also I have aimed at a certain unity of topic and argument, a unity indicated by the title of the volume. A brief summary may help the reader to grasp that unity and to follow the somewhat scattered argument. Man, I contend, is more than a machine, and more than a mirror that reflects the world about him. He is an active being with power to direct his strivings towards ideal goals; and there is ground for belief that those goals are neither wholly illusory nor wholly unattainable. There is no novelty about this view; but there is novelty in the argument by which the conclusion is reached. The same view has been propounded a thousand times by that form of wishful thinking which is commonly called philosophical. In this case the conclusion has been forced by the pressure of the evidence during more than forty years of cold and sceptical inquiry. The process is indicated in briefest outline in the first three essays of this volume. Any reader who may desire to follow the process in more detail may turn to my various published works, more especially to my *Body and Mind*, which remains pivotal for all my later thinking.

In spite of perennial discussion sustained through twenty-three centuries, we have but little understanding of man's nature, his powers and potentialities. Science is only now beginning to attack these problems with some slight promise of success. All advance along this line has been hampered and retarded: on the one hand, by religion, which has regarded man's nature as a sacred mystery not to be profaned by the prying fingers of science; on the other hand, by the obstinate prejudice that man is a machine, that he may be interpreted and explained without remainder in terms of the principles of physical science, a corollary from the larger prejudice that the whole universe is but a mechanical system. Our ignorance of man's nature has prevented, and still prevents, the development of all the social sciences. Such sciences are the crying need of our time; for lack of them our civilization is threatened gravely with decay and, perhaps, complete collapse.

Such understanding is not in principle impossible. If we can but achieve it in some considerable measure, we may fairly hope that the life of the human race may be vastly improved.

One step of advance we have made: namely, we have learnt that it is not true that 'all men are created equal', in the sense that they are of equal natural endowments;

rather, some men and some races of men are more highly endowed than others—have reached a higher level in the evolutionary scale. Human nature is not uniform, and is not established unchangeably at any one level. It is capable of further advance and very susceptible to deterioration. If we had some guarantee against widespread deterioration, our race might face the future hopefully. If we knew how to ensure some general biological advance, our hope might rise to confidence. But if some general deterioration of human qualities is inevitable, man's future must be dark and dreadful. This is the first lesson taught us by the application of the scientific method to the problems of human life. Can we profit by this new insight? Can we turn it to good account? Can we use our little spark of intelligence and our germ of purposeful spontaneity to turn the tide that seems to be setting strongly against us? That is the great question.

Critics will say that the tone of this volume is pessimistic. But that will be unjust. I am constitutionally optimistic; and if these essays strike a sombre note, that is the consequence of my lifelong studies of man, his powers, his efforts, his successes, his failures, his follies, his crimes, and his gleams of nobility. Man is a feeble and fallible creature; and he is in a most difficult and dangerous situation. But escape is not impossible; though prayer and fasting will not suffice. Man is not the passive sport of physical forces. He has, in germ at least, powers of understanding and of action which may yet turn the scale in his favour and justify the ways of God to man.

I add a few special words concerning some of these essays. All the essays here included (with the exception of the ninth) have been written since my fiftieth birthday and represent my matured convictions. I have found that, for the most part, my scientific friends refuse to take the first essay seriously. They insinuate that it was written with some private end in view; or they infer that I am the victim of some strong religious bias acquired in childhood, and hitherto pretty successfully concealed. It is of no avail to protest that I am honest in intention; that I here conceal nothing and make no pretence. But it may be worth while to state bluntly that I am not, and never have been, a member of any church; that for nearly fifty years I have weighed the evidences and the arguments in an impartial and agnostic spirit and have found that they incline me more and more towards the vague though positive conclusion here indicated.

The position represented in the second essay is also the outcome of much slowly maturing thought. At least twenty years of intensive study were required before I could clearly see the central importance of this problem for the theory of man—could see how the failure to grasp it, to define it, to take a decided stand on the side of the fundamental nature of purposive action, had sterilized well-nigh all the psychology of the nineteenth century and most of that of the twentieth. And though I have read and written and thought much about the problem, it is only recently that I have been able to map it out clearly, as in the present essay.

The discerning reader will see that the more popularly written essay, 'Was Darwin Wrong?' is intimately related to the larger questions discussed in the opening essays of the volume. When I was an undergraduate at Cambridge, neo-Darwinism was coming into fashion, and I accepted it somewhat superficially. But I was not satisfied. It seemed to me that the rejection of the Lamarckian theory was in the main a corollary of scientific materialism. I made some unsuccessful attempts to initiate a co-operative endowed experiment for the testing of the question; for it seemed to me too large a job for any one man to tackle. Many years later, in the course of my reading in preparation for my *Group Mind*, I became impressed anew with the need of a definite answer, positive or negative, to the Lamarckian question. For such an answer was urgently required, not only by biological theorists, but also by

viii

historians and all students of social problems, theoretical
and practical; at many points the interpretation of historical
and social facts was at a deadlock for lack of such knowledge. Although I was close to my fiftieth birthday, I
decided to undertake single-handed the difficult and
laborious task, giving to it what time and energy could be
spared by the head of a large department in full teaching
work. I initiated the experiment mentioned in the essay.
Harvard and Duke universities have generously supported
my effort; and the results hitherto achieved have surpassed
my most sanguine expectations. At the time this essay
was written the experiment was in its seventh year only;
now it has entered on its fourteenth year, and the indications of a positive answer to the Lamarckian question have
become very much stronger in many ways. Nevertheless,
the general argument remains unchanged, and I have merely
brought the essay up to date by altering a few words and
figures.

For the two essays on psychical research I offer no
apology. They sufficiently express my attitude to that
field. But, in accordance with my new policy of blunt
outspeaking, I will add this: if, like most of the professed
students of the nature of man, I had failed to give any
support or co-operation to the small band of earnest 'psychic
researchers', I should hang my head in shame. Thanks to
the liberal-minded attitude of Duke University and of its
President, I have succeeded in giving effect to the policy
advocated in the second of these two essays. And I now
point with pride to the most important of the first fruits
of that policy, namely, a monograph on *Extra-sensory
Perception* by my friend and colleague, Dr. J. B. Rhine,
to be published almost simultaneously with this volume,
by the Boston Society for Psychic Research (of which
society I am one of three or four co-founders). The monograph will speak for itself. Here I will say that, in my
considered opinion, it will establish the reality of both
telepathy and clairvoyance and, for the first time, bring

them definitely into the field of recognized and approved experimental science. I will add that this work of Dr. Rhine, together with the positive results of my Lamarckian experiment (in which he has co-operated of late years)—that these two pieces of work from the same laboratory will be found to have given to biological materialism the heaviest blow that it has suffered hitherto.

As for the essays on eugenic topics, they will inevitably antagonize a considerable proportion of those who dip into this volume. Favourable consideration of the Lamarckian theory excepted, there is nothing more injurious to the reputation (both popular and in the scientific world) of a man of science than to be mixed up with psychic research, unless it be a display of keen interest in the eugenic problem[1]. I have suffered much under both heads in the way of loss of reputation, unpopularity, slanderous misrepresentation, and scornful hostility. It seems to be my fate to espouse unpopular causes; but to support them so temperately and with so much critical reserve that I am as little acceptable to the minority in opposition as to the dominant crowd. My only recourse is to find what satisfaction I may in the 'bloody but unbowed' attitude, and in the approval of a select few.

During the last forty years I have published a long array of scientific hypotheses, and a number of suggestions for social and political improvements. Not one of these has attained to wide acceptance. It is therefore very gratifying that my suggestion of family allowances as a eugenic measure, after being ignored for twenty-seven years, is finding influential approval. At this last moment before going to press I learn that that eminent authority, Prof. S. J. Holmes, in a newly published book, gives it his powerful support. I am encouraged almost to hope that at some remote date my Eugenia scheme may be realized by some enlightened

[1] Unfortunately, the converse proposition is equally true: nothing so easily establishes a biologist in popular esteem as a scornful attack on eugenics and eugenists!

multi-millionaire. For the scheme, though it is put forward in a playful form, is very seriously meant. I am convinced that no rich man could devise any better method of applying his wealth to the lasting benefit of mankind. Since the writing of this essay I have taken part in the rapid creation of a great university. This experience makes the realization of the Eugenia project a less difficult and less remote possibility than it had previously seemed to be.

I conclude by pointing out that, just as the twelfth and the fourteenth essays are condensations of books already published, several of the others, especially the first, the third, the sixth, and the last, are merely preliminary sketches of books still to be written.

WM. McD.

DUKE UNIVERSITY, N.C.
February 1934

CONTENTS

RELIGION AND
THE SCIENCES OF LIFE

I

RELIGION AND THE SCIENCES OF LIFE[1]

THE sciences of life are widely reputed to be dangerous to religious belief in a higher degree than the physical sciences; of all the sciences of life, psychology is, perhaps, most open to this reproach. It may, therefore, be of some general interest if I, who have devoted more than forty years to these sciences, testify, while still pre-senile, that these prolonged studies have led me to a position more favourable to religion than that from which I set out. They have, in fact, led me from agnosticism to religion. Growing up in the midst of the great evolutionary controversy, a keen reader of Lyell, Darwin, Spencer, and Huxley, I had reached, while still a mere boy, a thoroughly sceptical and agnostic position. I had become 'a pagan suckled in a creed outworn'. From that position I have very gradually advanced (or, according to the taste of the reader, receded) to a more liberal one. Since this change has been brought about, to the best of my judgment, mainly by my scientific studies, it may be worth while to indicate very concisely the lines of evidence and reflection that seem to have played the chief part.

Deserving of the first place in this record of a sceptic's progress is my increasingly vivid realization that, in spite of all the splendid achievements of modern science, we still live surrounded on every hand by mysteries. On the frontiers of science we look out into infinite distances where all is dim and uncertain, where there loom up questions with which we vainly struggle.

Equally conducive to caution and humility is the mutability of scientific theories. I have seen many theories

[1] Reprinted from *The South Atlantic Quarterly*, January 1932.

I I

come and go. I have seen the answers of science which seemed satisfying and final to one generation thrown by the next to the scrap-heap of exploded fallacies. Need I cite instances? Out of many I will mention three only. First, the theory of strict determination of all events, in the sense that excludes creativeness and volition. A generation ago this theory was taught by men of science and philosophers alike with a dogmatic confidence that the Pope of Rome himself might have envied. To-day it is gone, undermined, disreputable, shown to be nothing more than an irrational prejudice, or at best, a methodological assumption.

Secondly, the theory of the adequacy of natural selection to account mechanically for organic evolution, for every adaptation and every appearance of purposive activity— a theory denying by implication all other agency and influence than the mechanical. That also has gone the way of exploded dogmas.

Thirdly, the principles of the great Newton, for generations accepted as the very foundations of all science; these also now belong to the past.

We may say of scientific theory in general:

> and anon,
> Like snow upon the desert's dusty face
> Resting a little hour or two—'tis gone.

It is the negations of science, rather than its positive affirmations, that have this fleeting and unstable quality. It is very hard to prove a negative; and I have learned that science, in making any general assertion, needs to be far more cautious than it has been in order to avoid including in its affirmations any implication of negations. It is in the fields of biology and psychology that a modest reserve in formulating negations is most imperative. In these fields science is still stumbling in the dark. In them the forces of science, unlike those of the physical sciences which march forward with at least an appearance of a united front, are divided into acutely opposed schools, into warring sects which cannot agree upon a common answer to any one question of the first importance, that is to say, to any question that goes beyond the province of mere description. In these sciences there are no explanatory hypotheses on which authorities are united. The great facts of evolution, reproduction, development, heredity, fixity of type, memory, purposive self-direction, learning or profiting by experience, intelligent adaptation, each stands with a great question mark beside it. And to each of these great questions science can return no answer. How, then, should it presume to issue negations?

But scientific caution and humility are not enough to lead the way to religion. They may keep the road open, but they cannot lead us along it. A certain boldness also is required, a readiness to grasp a vast range of converging evidence, each item of which, standing alone, can lead us nowhere.

The late poet laureate of England said, 'Man is a spiritual being; the proper work of his mind is to interpret the world according to his highest nature, to conquer the material aspects of the world so as to bring them into subjection to the spirit.'

Those words seem to me to define very simply religion and the whole duty of man. Yet simple as is this statement, it is not wholly clear until we know what Robert Bridges meant by saying that man is a 'spiritual being'. In the light of many talks which I had with him and of personal knowledge of him gained from our friendship, I feel sure that fundamentally he meant in those words to state a truth that results directly from observation and experience, a fact independent of all special revelation and tradition, an empirical fact as well established, as indisputable, as any you may find stated in the most authoritative text-book of physical science; the fact, namely, that it is the nature of man to recognize the true, the good, and the beautiful, to esteem highly all such things, to aspire towards them, to strive to preserve, augment, and create truth, goodness, and beauty.

Yet in calling a man a spiritual being Bridges meant more than the statement of this empirical fact of observation. He meant also to imply, as in the next sentence he actually asserted, another truth, one which is not ascertainable by observation, but which yet is established by universal assent; the truth, namely, that these capacities of which I speak are the highest part or aspect of man's nature. There we have a truth of a different order from any of the truths of science, a truth achieved by a process entirely different from any employed by science, and yet as indisputable as any science can boast—and far more important. This truth, that what we call spiritual in man is the highest part of his nature, is reached by way of a judgment of value. Such judgments are intuitive; unlike the hypotheses of science, their affirmations cannot be tested, proved, or verified by any observation or experiment. They have no place in the processes of scientific discovery; they belong to a different sphere; and yet this particular judgment stands firmly established by the universal assent of mankind. Such judgments were thus established long before science began to take shape, and they will stand firm, we may safely say, when most, perhaps all, the facts and theories of the science of our day shall have been rejected as false or transformed beyond recognition.

Not only are these affirmations of the supreme values arrived at by processes entirely other than those which science employs; they are independent of science in the sense that no conceivable discoveries made by scientific methods can refute or shake them.

The words of the poet which I have cited go further than the affirmation of man's spiritual nature and the supreme value of the spiritual. They assert also that it is the proper work of man to conquer the material aspects of the world and to bring them into subjection to the spirit. In making this assertion the poet implies that in some measure, however slight, man has the power thus to subjugate the material aspects of the world, to make those aspects subservient to his spiritual values, to transform the world about him, to 'remould it nearer to the heart's desire'. The same words imply also that man can in some measure remould his own nature, can develop and enrich his spiritual side, while making subservient to it the processes of his material frame.

Now this implication is essential to religion; but unlike the affirmation of the supreme value of the spiritual, it can be and has been attacked and denied by science. Science cannot impugn the affirmation of the supreme value of the spiritual; but it may deny, and through the mouths of many of its leaders it has denied, that the spiritual is of any effect in the life of man. Science cannot deny that man recognizes and acclaims truth, goodness, and beauty in all their forms; but it has denied that man's aspiration to conserve and to create these values is of any efficacy. And this is the most fundamental part of the attack of science upon religion. For, if this denial is well founded, religion is wholly illusory.

But religion of the more positive kind goes further. It asserts that these spiritual values are not merely incidents in the experience of individual men, realized and achieved by them in their various degrees, feebly and dimly by the common man, more intensely and richly by the highly endowed and developed natures who represent the peaks of humanity. It assumes, at least as a working hypothesis, that, in these individual experiences, man does not merely go through peculiar phases of emotion, nor merely find in them the stimulus to strive effectively for the realization of spiritual values. It asserts rather that in such experiences man makes contact with an aspect of the universe that is real and supremely important, an aspect which takes precedence of the physical realm. Furthermore, religion assumes that he not only makes contact with this realm but also shares in it, partakes of it, is influenced by it, and in return can contribute something, however little, to it.

These, then, are the two great affirmations of religion which science disputes: first, the spiritual ideals of man have causal efficacy, they are potent to change for the better both man's own nature and the world in which he lives (this we may speak of conveniently as the affirmation of spiritual potency). In the second place, man, in so far as his spiritual nature is developed, can and does participate directly in the life of a realm of spirit infinitely surpassing in extent and power his own small spiritual spark (this we may speak of as the affirmation of spiritual participation). These are the postulates of all religions, save the most tenuous. Religion claims that they are pragmatically verified by a wealth of human experience. The essence of the conflict between science and religion has been that science has seemed to deny these two affirmations. And it is notorious that at the present time biology and psychology are widely regarded as more active and positive in this denial than the physical sciences.

As biologist and psychologist seeking the truth in the sphere of religion, my task has been to inquire without prejudice, first: Do these sciences truly afford ground for such denials? Secondly, do they not rather, when largely considered, afford positive support for these two fundamental affirmations of religion?

I observe that the very existence of science invalidates its own denial of the postulate of spiritual potency. Science itself is a magnificent monument testifying to the efficacy of man's spiritual ideals, especially his ideal aspiration after truth. This point might be developed at great length: but I forbear. I turn to the specific allegations on which the denials are founded. They are the double allegation that man, with all his wonderful powers of intellect and spirit, is a mechanistic product of a mechanistic evolution.

Now it seems to me abundantly clear that neither of these allegations is well founded or even respectably supported. It has become almost a matter of general agreement among biologists that the mechanical theory of evolution, of which the principle of natural selection was the backbone, has broken down. On all hands we find biologists either accepting the significant expression, 'creative evolution', or speaking of orthogenesis, which means much the same thing, an evolution directed towards a goal. Under these terms the admission is made that mind, instead of being a mere product or by-product of evolution, has been and is, in some sense and manner, the essential active agent in evolution. For the only directive and creative agency we know or can conceive is of the nature of mind.

So recently as the last week of September in London this battle was fought over again by leading biologists; the progressive, large-minded biologists were arrayed against a few ultra-conservative reactionaries, survivals from the nineteenth century, too old and too set in their ways of thinking to desert the dogmas of their youth.

We have, then, as the progressive leaders of biology frankly recognize, no adequate theory of the evolutionary process; yet the fact of organic evolution is one evidence of the primacy of mind in the universe and of its leading role in the world drama. And it is no less clear that, if the race of man is to make further evolutionary progress, such progress can come only through the effective working of his spiritual ideals.

When we turn to the biology of existing organisms, to the facts of their self-regulation, maintenance, and repair, and especially to the processes of reproduction with all the phenomena of heredity, we again find the mechanistic theory hopelessly out of court, with an increasing recognition of that fact among biologists. Everything happens as though regulated for the sake of one great end, the maintenance of the individual and the perpetuation of the type. And the more intimate our acquaintance with these processes becomes, the more clearly does this appear. Dr. J. S. Haldane, one of the greatest of living physiologists, recently asserted bluntly: 'Physical science cannot express or describe biological phenomena, so that its claim to represent objective reality cannot be admitted.'[1] And he insisted afresh on the simple truth that, since the processes and entities described by physical science are abstractions (as, for instance, mathematical formulae deal only with abstractions such as numbers and space), the principles formulated about these abstractions by physical science have no valid place in the more concrete science of biology.

But it is when we turn to the still more concrete phenomena dealt with by psychology that the inadequacy of physical principles becomes most obvious and indisputable. The mechanical psychology of the nineteenth century, in spite of the efforts of the 'behaviourists' to revive it, is utterly bankrupt. Here we see the importance of the principle that natural events must be interpreted in the light of their most developed and complete forms. It was only by confining their attention to the lowliest manifestations of mind and to partial and abstract aspects of its activities that the mechanical psychologists succeeded in making their dogmas seem plausible.

When we consider the larger and higher activities of man, it is as clear as daylight that those activities conform to laws quite other than the laws of physics. As hitherto formulated, the laws of the physical world are mechanistic (which means that the future course of events is wholly determined by the present and the present by the past) and, therefore, non-creative. This remains true no matter how subtle, immaterial, vague, and amorphous the entities, substances, or ultimate postulates of the modern physicist. The activities of men, on the contrary, are purposive; they conform to teleological laws and are ʳcreative in the

[1] *Philosophical Basis of Biology* (London 1931).

fullest sense. Especially is it clear that man's higher activities are prompted and sustained by spiritual ideals, by his aspirations towards truth, goodness, and beauty. It is ridiculous that it should be necessary to point to, and reaffirm, such obvious and indisputable facts. Yet the science of the nineteenth century was almost quite blind to them; while the reactionaries of to-day still cling wilfully to that blindness, acclaiming it a virtue. Their position is pathetic in that, whereas the belief in the mechanistic determination of human life was deduced from certain principles of physical science, the physical scientists themselves have now abandoned those principles in their own sphere, while the reactionary biologists and psychologists remain clinging to the unsupported dogma like sailors clinging desperately to the mast of a sinking ship deserted by its officers.

Let us notice one anti-religious argument of a different kind which calls for special attention by reason of the eminence of its most recent exponent. Professor Sigmund Freud, world-famous author of the psycho-analytic doctrines, has recently published a book called *The Future of an Illusion*, in which he claims to show that all religion is illusory. What is his argument? Merely this, that the nature of man is such that the race inevitably acquires religious beliefs. Strange argument! The same fact has frequently been used as the surest evidence of the truth of religion; as when Descartes held that the idea of God is innate in the race and that, therefore, theism is true. Now I do not for a moment accept Professor Freud's fantastic theory of the origin of the belief in God. But I do agree with him that the nature of man is such that he develops religious beliefs. The fact is obvious. If it were not so, man would not acquire religion, no matter how true its doctrines nor how obvious the evidences of them. I merely point out that Freud's argument in that book is a complete *non sequitur*. Its premises point at least as strongly to the opposite conclusion. Freud's book, his famous attack on religion, is but another illustration of the fact that man's intellect is a feeble thing, liable, even in the greatest men, to be led astray by emotional bias and by prejudices unrecognized by the thinker.

I must not linger on the first of my two questions: Do biology and psychology render untenable the fundamental postulates of religion? The answer is clear: they do not. I must pass on to the second and more difficult question: Do they afford positive support to religion?

I have already in part answered this question. These sciences do show that the living being is more than a concatenation of physical forces; they do support most fully the affirmation of Robert Bridges with which I opened this article, the affirmation of spiritual potency, the affirmation that man is a spiritual being whose proper work is to make the spiritual prevail over the material or physical aspects of the world. They show us mind active all along the scale of life, becoming increasingly effective and predominant until in man spiritual ideals become capable of transforming the world, or at least promise such transformation.

But do these sciences afford positive support to that further affirmation of all positive religion, the affirmation of spiritual participation? Do they yield us any evidence that the spiritual is not confined to the small centres of individual consciousness that we call men, but that rather these individual sparks of spirituality are parts, or fragmentary expressions, of a spiritual realm that far transcends them and is the most fundamental, the predominating, the primary aspect of reality? If these sciences yield such evidences, it is as much as we can ask or hope of them. We must not expect of them specific support for any particular creed or theological doctrine. That is for philosophy and theology.

Suppose, for the sake of argument, that we accept for the moment the Darwinian or, rather, the neo-Darwinian account of organic evolution, untenable as it is. Even under this scheme the theory of evolution postulates the *struggle for life*. Animals do not merely react mechanically to physical impressions—they struggle to survive, to hand on the torch of life; they struggle for more and better life. Their struggle is a series of activities which, though lowly and relatively simple, are yet allied to and are of the same fundamental nature as our own purposive actions, the purposive powers which reach their fullest expression in the spiritual activities of men, in their acts of creative will inspired and guided by spiritual ideals.

Now a purposive action, when considered in isolation, is strictly speaking unintelligible; it has not the intelligibility of an isolated mechanical event, such as the impact of one billiard ball upon another. It is for this very reason that it is so hard to persuade many psychologists that even human activities are truly purposive. They cannot see through and comprehend the isolated purposive act from beginning to end. That difficulty I admit. I insist upon it. But I do not for that reason resort to the absurdity of denying the obvious facts. There is nothing more obstructive to the advance of knowledge than a certain unformulated dogma implicitly accepted by many men of science, namely, the dogma that what we cannot fully understand cannot happen. We cannot too strongly insist that the bounds of the possible do not coincide with and are not set by the limits of our present powers of comprehension.

I submit, then, that every instance of purposive activity, whether human action of the most exalted type or the simple striving for life of a lowly animal, points beyond itself to a larger purpose of which it is but a momentary and fragmentary expression. Here we have one of the evidences of the view, often asserted, that all life is one, that all living creatures are but twigs upon the single tree of life through which runs one common stream, a stream of purposive activity; and, since in man this stream rises to the level of spiritual activity, we may infer that the common stream is one of spiritual activity also, however partial and slight its more lowly expressions may be.

But man's activities are not only purposive. Also they are sometimes and in some degree logical. And we find that the more logical our activity, the more successfully does it cope with the world about us; that is to say, there is a certain congruity between our logical activities and the world to which we apply them. A great question confronts us: How has that congruity arisen? Two answers are current. The one asserts that the congruity exists because man's nature has been moulded and adapted to cope with the physical world; that, for this reason alone, the laws which his mind obeys are congruous with the laws of the physical world. The other answer is that the laws of reason are primary and fundamental; that they are not the products of an evolutionary process, but are intrinsic in the nature of mind or spirit. If this be so, it follows that the congruity between them and the laws of the physical world can only be interpreted as meaning that the physical world has been shaped by and is an expression of spirit. This is a momentous conclusion, but it is one to which leading physicists are themselves inclining. What has psychology to say on this great issue?

I will present only one argument. The logical powers of the human mind find their highest expression in works of mathematical genius. Now such genius appears sporadically. It is in some cases manifested by mere children (as by the boy Blaise Pascal) and, by discipline, becomes full-blown. In other children it is manifested at an early age to an astonishing degree, and then fades away. Now I say: It is inconceivable that this logical power thus mysteriously manifested is the product of the processes of organic evolution, of the mere struggle of living things to maintain life. It must be that it pre-exists, that it is intrinsic in the nature of the universe and finds expression in various degrees in human life according as the human organism is fitted for the exercise of it.

The purposive and the logical nature of man's activity alike mark him as a vehicle or channel through which the spiritual realm finds partial expression.

But it is in man's power to appreciate and to create beauty that we find the clearest and most positive evidence of this view. Music has often been called the most spiritual of the arts; for it is least dependent upon or connected with any spatial or material representation. Now, as we know, the musical faculty is mysteriously correlated with the mathematical; and, like the latter, it sometimes is manifested in astonishing degree in young children, as in the young Mozart. As I said of mathematical genius, it is inconceivable that such musical faculty can be the mere by-product of a naturalistic evolution, of a struggle for life. Its degree goes far beyond any biological utility. And the same is true of all the richer instances of man's aesthetic powers, whether of appreciation only or of creation also. Can any one seriously maintain that the godlike powers of Beethoven or of Shakespeare can be validly conceived as created by a mere struggle for survival? As Professor Bergson once said to me, Shakespeare gives expression to much more than his individual experience accumulated by sympathetic observation of his fellow-men; at the least, his art expresses the experience of the race.

The same is true of all great poets and of all great poetry. The great poet feels himself to be the channel through which a spiritual activity other than his own finds expression. And every great poet implies that, when we are under the spell of beauty, especially of great art, we are in effective contact with a spiritual realm, however dimly it may be revealed to us in such experiences. I believe that the most sober psychological consideration of the facts must lead us to the conclusion that the poets are right.

Let us remember that all our descriptions are only symbolical, that our most sober describing of physical things is essentially a stringing together of symbolic sounds which convey our meaning only in so far as they are symbols of a reality that we most inadequately apprehend. And the language of art is no less and no more symbolical; whether it be a great cathedral, a poem in stone; a symphony, a poem in tones; or a poem in words. That which these things symbolize, or imperfectly shadow to us, is the realm of spiritual reality. And, if this is true of the beauty of art, it must be true no less of the beauty of nature. For, between the beauty made by man and the beauty of nature, there is no essential difference of kind. The beauty of nature, which also speaks to us in symbols, is symbolic of the spiritual nature of the universe. And this the poets have confidently asserted:

> Look! how the floor of heaven
> Is thick inlaid with patines of bright gold;
> There's not the smallest orb which thou behold'st
> But in his motion like an angel sings,
> Still quiring to the young-eyed cherubins;
> Such harmony is in immortal souls;
> But, whilst this muddy vesture of decay
> Doth grossly close it in, we cannot hear it.

These lines have been called the most sublime of all those written by our greatest poet. They are less exact than the descriptions of the astronomers, but not for that reason less true. They are written in a different system of symbols.

We are confronting a double mystery: on the one hand, the beauty of nature; on the other, man's susceptibility to it, his capacity to respond to it, to appreciate it, to be lifted up by it and, thus stimulated, to augment by his own efforts the beauty of the world. It is a congruity, a harmony, between man's mind and the world about him similar to that which we have noted in the logical sphere; but this harmony is the more convincing of the two. And if the sceptic asserts that the world contains no beauty in itself, that it merely appears beautiful to man, and that this is one of man's many illusions, we reply by asking: By what naturalistic process of evolution could this strange power of man have been engendered, the power to see and feel beauty all about him and to create new beauty?

No, the poet is right when he speaks of

> . . . those obstinate questionings
> Of sense and outward things,
> Fallings from us, vanishings;
> Blank misgivings of a Creature
> Moving about in worlds not realized,
> High instincts before which our mortal Nature
> Did tremble like a guilty thing surprised.

And of

> . . . those first affections,
> Those shadowy recollections
> Which, be they what they may,
> Are yet the fountain-light of all our day,
> Are yet a master-light of all our seeing;
> Uphold us—cherish—and have power to make
> Our noisy years seem moments in the being
> Of the eternal Silence.

In the moments of contemplation of beauty the boundaries of our personalities are in some degree transcended, we come near again to the universal spirit which glows, feebly or strongly, in each one of us; we partake more completely of it, we are in some measure re-absorbed into it.

I will point only to one branch of psychological investigation where the strictest methods of empirical science yield evidence in support of the view to which our argument leads us. I mean that borderland region of rare and disputed phenomena known as the field of psychic research. We have made but very little progress in this difficult field. Men of science have for the most part shunned it, in fear disguised as contempt. Yet a few devoted students have made some slight entry, have established some facts ignored and, by implication, denied by orthodox scientists. And these observations suffice to establish on a solid scientific basis this truth, that each one of us, each individual centre of consciousness, is not completely shut up within a prison whose only windows are the sense-organs, as science has commonly asserted. There are other possibilities of communion of one spirit with another, other channels of communication of which we have but little understanding. It is as though the prisoner in the penthouse had not only the narrow windows of the sense-organs, but also a radio-apparatus which, if he can but learn to use it, may bring him tidings from the remotest corners of the world, surmounting the barriers of space and perhaps also those of time. Here, then, we have yet another line of evidence that each man is not what to so many scientists he has seemed to be, a fortuitous concatenation of physical forces, but is rather a ripple of the mighty ocean of spirit, an individualized ripple, small and feeble, yet sharing in the nature of the whole and not wholly detached from it.

Such is, in my considered opinion, the bearing of the most sober and impartial survey of biological and psychological science upon religion. The evidence supports the view that religion and science have ample scope to approach ever nearer to the truth without essential conflict, to achieve a fuller understanding of the spiritual aspect of reality without any sacrifice of the essentials of religion or any offence against the most rigid canons of scientific reasoning. It bids us be of good cheer: for truth and reason are not mere adaptations to environment; rather, truth is a spiritual ideal and must in the end prevail, and reason is of the essence of the universe; while beauty is the symbol of immortal things, a symbol which we can but dimly comprehend, things which the poet alone may tell of in the symbolic language of his art.

II

MECHANISM, PURPOSE, AND THE NEW
FREEDOM

THE problem of the relation between mechanism and purpose is of profound theoretical interest. It is the most fundamental of the great perennially disputed problems. And, unlike some other of the great unsolved problems, it is also of far-reaching and profound practical importance. The kind of answer we give to the question affects in a multitude of ways the conduct of our lives, the form and working of all our institutions, our science, our law, our politics, our ecomonics, our morals, our religion.

Men of all periods have made use of explanations of two types, in dealing with natural events have postulated causation of two very different kinds. It has often been said that the primitive animist explains all events in terms of purposive causation. But it is not true; the statement is a libel on the intelligence of the savage. It is true that he gives to purposive action a wider role than we do; but also he assumes mechanical causation in all ordinary practical activities, as when he builds a house or a boat, makes and wields his tools. And, at the other end of the scale, the nineteenth-century materialist, who in theory explained all events as the mechanical interplay of hard atoms, in practice recognized the reality of purposive causation in the conduct of his own life and in all dealings with his fellow-men; an inconsistency which reached its most amusing climax when T. H. Huxley, in his famous, Romanes Lecture, exhorted us to defy the purely mechanical universe in the interests of morality and human welfare; or when Bertrand Russell, after eloquently expounding, as the principles of a free man's worship, the stoic acceptance of a strictly mechanical universe, of which every event, including all human thinking and activity, is strictly determined in every slightest detail for all time, appeared as a social and moral reformer, an advocate of nudism, free love, and the general principle of doing as you please.

These modern stoics are no whit less inconsistent than the stoics of antiquity who, accepting a strictly determined mechanical universe, exhorted men to subdue their passions, control their natural impulses, and live according to reason. And, one may add, they are rather more inconsistent than the savage who in his treatment of his boat or his spear applies at one moment the mechanical, at another the purposive, principles of explanation and causation.

It is true that the development of science has brought a considerable restriction of the sphere of purposive causation and a corresponding extension of that of mechanism, with a more exact demarcation of the two spheres. The history of the process is familiar. It began with Democritus in the fifth century B.C., and culminated with the Darwinian revolution in biology. It was a series of bitter controversies in which, on the whole, the advocates of universal and exclusive mechanical causation gained ground, and in the end claimed, and to many minds seemed justly to claim, a sweeping victory.

Surveying this age-long controversy, the verdict must be, I think, that the defence of purpose was in the main badly conducted. It fell into the hands of metaphysicians and theologians; of the former the majority have capitulated to the attacking scientists; while the latter have continued the defence mainly on the ground of the evidences of design in the creation of the world. In this way there grew up the irrational groundless convention that science does and must recognize only mechanistic causation, while purposive causation is left to a minority of belated metaphysicians and to theologians defending a lost cause which they cannot afford to abandon. That brief statement characterizes, not unjustly I think, the state of this great issue in the last years of the nineteenth century.

I am concerned here to discuss the bearing on this question of the most recent developments of the sciences, and there can, I think, be no disputing the proposition that these developments make strongly for a revision of that *fin de siècle* position and for the re-establishment of purposive causation as an indispensable category of our thinking.

Before attempting such a brief survey, let me define the various ways in which purpose finds recognition in our discourse. Philosophers have discussed the problem of teleology as though teleology were a broader category than purpose, some making this claim explicitly. The most general evidence of teleology has always been the appearance of design in things; and appearance of design arises from the improbability that anything so complexly organized as the things in question should have arisen through the fortuituous concurrence of atoms or of physical energies working blindly, mechanistically.

But design implies purpose; therefore the Great Designer is assumed to have created the world with some such purpose as the realization of order, or harmony, or beauty. These two steps result in the Deistic theory which was so widely held by the men of science of the eighteenth century. Others went farther along this road and adopted the anthropocentric theory; yielding to the egotism natural to man, they argued that the order and harmony and beauty of the world were created in order that man should enjoy them.

For those who stopped at that position, the world of nature remained a purely mechanical system, all living creatures being machines of various degrees of complexity, and man merely the most complex machine of all, though a machine capable of contemplative enjoyment.

Such a theory may be called the theory of *initial teleology*; a better term, I think, than the one proposed by Professor Driesch, namely, 'static teleology'.[1]   The mechanistic scientist has no serious quarrel with this theory.   Such initial teleology seems to him harmless enough, for it leaves him a free field for the explanation and prediction of all events in mechanical terms.   But the theologian and the common man have never been contented by initial teleology.   The strong anthropocentric urge has carried them on to believe that the Creator has not abstained from all intervention with our world subsequently to the creative activity, to believe rather that He intervenes more or less to redirect its otherwise mechanical course.   Thus the Creator becomes also the Great Engineer; and we have an *external contemporary teleology*.   This exhibits three varieties: (*a*) that of the popular mind which sees the hand of God in all unusual events; (*b*) that which leaves the inorganic world to undisturbed mechanical causation, but sees the finger of God in the functioning of all living things; (*c*) that which restricts the teleological intervention to critical occasions in the mental life of men. All three beliefs are varieties of what may be called *interventional teleology*.

Another variety of teleology is that of the biologist who, seeing that many of the processes of living organisms (especially the processes of restitution of form and function after disturbance of the normal course) cannot be explained mechanically, postulates some unknown factor (utterly unspecified except as non-mechanical) which, steering the bodily processes towards their normal goal, keeps them true to the specific type.   This, the teleology of many vitalists, may be called *neutral teleology*.[2]

Now it is certain that all these varieties of teleology

---

[1] Cf. his *History of Vitalism*.

[2] In recent times it is represented by the 'entelechy' of the earlier writings of Professor Hans Driesch and by the special form of teleologically operating energy postulated by the late Eugenio Rignano.

derive from man's experience of his own purposive activities
and from his belief that such activities (both his own and
the similar activities of his fellow-men) are causally effica-
cious in the world of natural events. It is certain that
men, if they had no such immediate awareness of, and
familiarity with, their own purposive activities, would
never have conceived any teleological explanations, and
that all such terms as design, plan, goal, function, efficiency,
serviceableness, fitness, struggle, striving, effort, could
have no meaning and could never have been evolved;
and it is doubtful whether such terms as tendency, force,
energy, causation, or any words expressing any valuation,
whether positive or negative, could have meaning to a
creature, conscious, but mechanical or mechanistic in its
workings; a creature such as the biologists and psychologists
of the post-Darwinian period have commonly asserted man
to be.

It is certain that man has conceived the Creator and
Designer of the world after his own image, in the light of
his knowledge of himself. It is certain that, if the
purposiveness of human activity can be explained away
as an illusory appearance, teleology of any kind becomes a
groundless and utterly unjustified assumption, 'purpose'
a meaningless word, mere *flatus vocis*; while religion and
morality remain merely fanciful and illusory constructs
of man's 'mythopoeic faculty'. To postulate teleology
of any kind, anywhere and anywhen, while denying the
reality and efficacy of purposive activity in man, is
strictly absurd.[1]

I propose then to show very briefly, first, that recent
developments of science have undermined the common
objections of men of science to the full recognition of
the purposive activity of man; secondly, that other recent
developments provide increasingly strong grounds for such
recognition.

[1] Yet this seems to be the position of Dr. L. G. Henderson in
The Order of Nature, and of Dr. Joseph Needham in *Man a Machine*.

OBJECTIONS TO EXPLANATION IN TERMS OF PURPOSE
UNDERMINED BY RECENT DEVELOPMENTS OF SCIENCE

The most general objection arises from a widely prevalent neglect to distinguish between the various forms of teleology defined above. The man of science naturally, inevitably, and properly dislikes interventional teleology in all its three forms; for the good reason that to accept such teleology is to admit that an indefinable proportion of phenomena are supernatural, and, therefore, by their essential nature, beyond the reach of scientific investigation and explanation. If the Creator frequently intervenes in the course of natural events to send us rain or sunshine, how can there be a science of meteorology? If earthquakes are expressions of divine or diabolic wrath, how may we hope to foresee them and to take due precautions? Further, the partial and increasing success of efforts to build sciences of meteorology and seismology affords good empirical support to this natural repugnance. Conversely, on the other hand, the dislike of all attempts to create a science of psychology (still widely prevalent in theological circles) is due largely to the fact that such attempts seem to threaten the last stronghold of interventional teleology.

Interventional teleology, so dear to the theologians, is the one kind of teleology that is logically incompatible with the spirit of science. While the theologian pins his faith to its reality, the man of science equally makes it a prime article of his creed that such intervention does not occur. This is the real crux of the long conflict between science and religion.

Science has no quarrel with initial teleology. So long as teleological activity is restricted to the creation of our world, the course of nature since that remote event remains a fair field for scientific explanation. And that is enough for the impartial scientist. The history of thought, from the later stoics to Newton, Voltaire, Priestley, and in our own time, L. G. Henderson and Joseph Needham, shows how readily men of science may accept initial teleology. Why, then, are they so reluctant to recognize the causal efficacy of human purpose? Such recognition carries no more logical implication of interventional teleology, the natural bugbear of science, than does initial teleology?

One reason, psychologically effective though quite alogical, is that the recognition of human purpose, though it does not logically imply interventional teleology, does open the door to it as a possibility. Indeed, this possibility is the last stronghold of all religion in any vital and significant sense of the word. The reality of some intercourse between the individual human mind and some larger mind is a vital assumption of all effective religion; and such intercourse implies interventional teleology. Hence human purpose becomes the object both of the cold dislike and suspicion of the impartial scientists and of the iconoclastic fury of all those who harbour the anti-religious complex.

Closely allied with this most general but alogical objection to human purpose is another, equally influential and equally due to confused thinking. Science rightly seeks causal explanation of all phenomena; and the category of causation is the indispensable foundation of all science that seeks to go beyond the merely descriptive stage. Now there prevails among scientists the belief that all causation is synonymous with mechanistic determination. The history of the genesis and spread of this belief is a subtle psychological problem. It is clear that its prevalence is a modern development, one which is closely bound up with the Newtonian system and with the nineteenth-century delusion of the adequacy of strictly mechanical explanations of all physical phenomena. For under that system all causation was necessarily mechanical causation.

No doubt a considerable part in the genesis of this belief was played by Aristotle's account of causation in terms of causes of a number of different kinds, among which final causes were one kind. For the expression 'a final cause' was commonly applied to the result achieved, or to be achieved, by purposive action. And it was asked: How can the result, an event still in the future, play a part in the causation of a present event? The notion is plainly nonsensical. Therefore away with final causes and every trace of 'finalism'. A main feature of the revolt against Aristotle which initiated the modern period of thought was this repudiation of 'final causes' and the growing traditional dislike of all thinking tinged with 'finalism'. Such is the unfortunate influence of 'terminological inexactitudes' upon our poor weak minds.

That this belief in the identity of all causation with mechanistic determination is still widely prevalent is shown not only by its explicit utterance by workers in many branches of science, but also by many implied assumptions of its truth. The assumption is implicit, for example, in the proposal of a numerous group of contemporary German psychologists to construct alongside of the established academic psychology (which they call 'scientific causal psychology') a psychology of a very different kind to which they would give the name *verstehende Psychologie*.[1]

Whatever the exact history of this belief, it is clear that it is without logical justification. Though we may regret the language in which Aristotle discussed causation, we must follow him in recognizing, in principle, causation of two different types. On the one hand, we may properly conceive of causal processes which involve no reference of any kind to future happenings. All such events as are capable of explanation in terms of antecedent and contemporary events involve causation of this kind only. Such events and such explanations, even though not conceived in strictly mechanical terms, are conveniently called 'mechanistic'.[2] But we are perfectly familiar with

[1] Jung, Münsterberg, Erisman, Spranger.
[2] As I have pointed out in my *Modern Materialism*, the failure of universal mechanical explanation leaves us with no other meaning or possibility of definition of the word 'mechanistic', a fact which

events of a type in which reference to the future seems to play an essential role, namely, all our deliberately planned actions, every action in which we achieve some result which we have first conceived as a possibility of the future and have desired and striven to bring about or realize. All such successful actions are clear instances of purposive causation. They are the only kind of teleological event of which we have direct and intimate knowledge. To put aside such causation as not truly causation on any *a priori* ground is a perfectly arbitrary procedure; it is to be metaphysical in the worst sense. Yet it is just what is done by a multitude of men of science who profess the utmost scorn for metaphysics.

seems to be recognized by very few of those who use the word. Such negative definition is not very satisfactory; and those who seek to find a more positive definition of the term 'mechanistic explanation' commonly offer some such phrase as 'the kind of explanation given by physics and chemistry'.

The mechanist position (as here distinguished from the strictly mechanical) is in high favour with the biologists especially. All the many adherents of what is called the organismic view belong to this group; all those who insist on the unity and wholeness of the organism; and all the many who have accepted the principle of 'emergence' as the key to hitherto incomprehensible properties of organisms. Here must be reckoned also most of the psychologists of the now so flourishing *Gestalt* school.

It is strange that no one hitherto seems to have attempted the synthesis of the kindred principles of emergence and *Gestalt*. The obstacle to such synthesis seems to be the reluctance of the leaders of the *Gestalt* school to commit themselves to the recognition of the causal efficacy of psychical process, a recognition freely made by many of the emergentists: H. S. Jennings, W. M. Wheeler, and especially the neurologist C. J. Herrick. Professor Lloyd Morgan, the arch-emergentist, hedged deplorably on this question, as I have pointed out in my *Modern Materialism*.

But to recognize the unity of the organism and even to suggest an explanation of the fact of unity by regarding each organism as one vast *Gestalt* or configurational field of energy, is not to solve all the problems of biology. Especially, all the facts of restitution of form and function after distortion, and, in only less obvious degree, all the facts of heredity and morphogenesis, remain recalcitrant to all mechanistic explanation, even when the organismic, the emergent, and the configurational principles are combined. As so rightly insisted by Professor Hans Driesch and other vitalists, they point to some teleological factor in the life of organisms.

The question whether these two kinds of causation are ultimately distinct, or whether either of them may ultimately be shown to be a disguised form of the other, and, if so, which one is the more fundamental, which the illusory—that question is one which only the progress of science far beyond its present stage can answer. The hasty dogmatism, the gross confusion of thought, and the sheer unthinking ignorance with which this vital distinction is commonly slurred over, or obscured in favour of universal mechanism, is the greatest blot on modern science.

The confused thinking which identifies all causal explanation with mechanistic causation and puts aside purposive interpretation as non-causal is an instance of what the sociologists call 'cultural lag'. Such identification seemed well founded and justified so long as the theory of strictly mechanical causation of all physical phenomena seemed tenable. But the rejection of that theory destroys all ground of such identification; and the continuance of that identification in the minds of so many would seem to imply a failure fully to accept or to understand the completeness of the breakdown of that theory. This lag would seem, then, to find its psychological explanation as an instance of wishful thinking, thinking motivated by a subconscious desire to retain a mode of thinking so congenial to the constitution of our minds. Professor Bergson has rightly insisted on this congeniality, and leading physicists have not been slow to follow him. But the completeness of the renunciation of strictly mechanical explanation and the new freedom which it gives to our thinking require to be emphasized.

All the world knows that of late years physicists have recanted much of their dogmatism, especially their dogmatic denials founded on the clear-cut distinctions of the Newtonian system, time, space, mass, momentum. From the point of view of our topic the great change has been the admission that the strictly mechanical system of explanation, explanation in terms of the kinetic causation of moving masses, was hopelessly inadequate even in the sphere of pure physics. A second change, hardly less significant, is the breakdown of the clear distinction between matter and energy. These two closely allied changes bring with them the abandonment of the dogmas of conservation of mass, of momentum, and of energy. A third great change is the substitution of statistical probability for strict predictability and the abandonment of the dogma of strict determination of all events.

In brief, the great change, the great renunciation is the frank admission that the strictly mechanical system of explanation is untenable.

Of course, it always was untenable. The change is a giving up of a convention which may have been useful in a limited way, but which was made to seem generally true and to be the solid ground of various dogmatic denials only by obstinately turning a blind eye towards many natural phenomena. Einstein has not really changed the nature of the universe, as the language of many physicists might seem to imply. He has changed only the physicist's ways of thinking about the world, has shown the inadequacy of a way of trying to explain natural phenomena which for some centuries had been conventional among physicists, a way which had been even more fanatically followed by the many philosophers and the many students of other branches of science (especially the biologists and the social scientists) who have been unduly dominated by the immense prestige of the physicists. Many of these still cling pathetically to the negative implications of the abandoned mechanical theory, like drowning sailors to a fragment of wreckage.[1]

[1] It was surely always obvious to the more impartial thinkers that the strictly mechanical system of explanation was not applicable to such facts as the chemical affinities and properties of elements and compounds. When I was an undergraduate student of chemistry we used to think of the atoms of a molecule as linked together with little hooks which were represented on the blackboard by

The great renunciation, the repudiation of the strictly mechanical system of explanation as adequate to all events, is now thorough and complete. But among the physicists (as among other thinkers also) we must distinguish two groups, namely: on the one hand, those who recognize that the great renunciation leaves room in nature for the recognition of the causal efficacy of purposive activities; on the other, those who still reject purposive activities as anathema, as outside the realm of nature, as supernatural or non-natural events and therefore in some sense unreal.[1] The scientists of the latter group

short straight lines. But no one, I suppose, took these mechanical symbols very literally; and we talked about forces of attraction and repulsion exerted by the atoms upon one another, forces which were supposed to act very powerfully across small distances. I remember also that in the pre-relativity and pre-quantum period a physicist of high standing, Wilhelm Ostwald, secured widely diffused interest in his proposal to overcome scientific materialism by abolishing matter altogether, substituting for it a great variety of forms of energy, among which were mental or psychical energies (*Vorlesungen über Naturphilosophie*, Leipzig, 1902).

'I believe,' writes a modern physicist, 'many will discover in themselves a longing for mechanical explanation which has all the tenacity of original sin. The discovery of such a desire need not occasion any particular alarm, because it is easy to see how the demand for this sort of explanation has had its origin in the enormous preponderance of the mechanical in our physical experience. But nevertheless, just as the old monks struggled to subdue the flesh, so must the physicist struggle to subdue this sometimes nearly irresistible, but perfectly unjustifiable, desire. One of the large purposes of this exposition will be attained if it carries the conviction that this longing is unjustifiable' (Professor P. W. Bridgman in *The Logic of Modern Physics*, New York, 1928). It may be added that the psychological root of this illegitimate longing is the fact that all our own causal activity directed upon the world about us, both animate and inanimate, seems to be (with certain possible but rare exceptions) exerted by moving parts of our bodies and, by means of such movements, communicating motion to other things. If it were common experience to communicate with our fellows telepathically or to ignite fire or to initiate other chemical changes by direct volition, the mechanical system would never have enjoyed exclusive favour. The reading of Professor Bridgman's book strongly suggests the need for a companion volume on the psychology of modern physics.

[1] Prominent among the former group (the smaller, I imagine) are Eddington and Jeans. The latter writes: 'The fact that "loose

(which includes many biologists and psychologists) incline to be scornful and resentful of the former, as of colleagues who have abandoned to the enemy the very citadel of science. Although they admit (under compulsion) the inadequacy of strictly mechanical explanations (explanations in terms of the momentum and impact of masses), and though they recognize that perduring particles of matter in motion are not necessarily assumed in all explanations of events, they still stoutly repudiate purposive or teleological activity as a natural process having

jointedness" [William James's favourite appellation] of any type whatever pervades the whole universe destroys the case for absolutely strict causation, this latter being the characteristic of perfectly fitting machinery.' And: 'Although we are still far from any positive knowledge, it seems possible that there may be some factor, for which we have so far found no better name than fate [why not 'purposive activity'?] operating in nature to neutralize the cast-iron inevitability of the old law of causation. The future may not be as unalterably determined by the past as we used to think.' Again, after showing that a mechanical universal ether is impossible, he writes: 'We are compelled to start afresh. Our difficulties have all arisen from our initial assumption that everything in nature, and waves of light in particular, admitted of mechanical explanation; we tried in brief to treat the universe as a huge machine. As this had led us into a wrong path, we must look for some other guiding principle.' And yet again: 'The picture of the universe presented by the new physics contains more room than did the old mechanical picture for life and consciousness to exist within the picture itself, together with the attributes which we commonly associate with them, such as free-will and the capacity to make the universe in some small degree different by our presence. For, for aught we know, or for aught that the new science can say to the contrary, the gods which play the part of fate to the atoms of our brains may be our own minds. Through these atoms our minds [our purposive strivings] may perchance affect the motions of our bodies and so the state of the world around us. To-day science can no longer shut the door on this possibility; she has no longer any unanswerable arguments to bring against our innate conviction of free-will.' To which it may be added that she never had any such arguments, a fact which seems to escape both Eddington and Jeans. These passages are cited from *The Mysterious Universe*, New York, 1930

Similarly Eddington says in a recent address: 'His (the physicist) first step should be to make clear that he no longer holds the position occupied for so long, of chief advocate for determinism, and that if there is any deterministic law in the physical universe, he is unaware of it (Presidential Address to the Mathematical Association, 1932).

causal efficacy. And even such a chemist as L. G. Henderson, who elaborates a powerful argument for initial teleology (for design and purposive direction in the creation of our world), continues to deny all causal efficacy to purposive activity in our contemporary world.[1]

Such repudiation of the strictly mechanical explanation, combined with refusal to recognize the efficacy of purposive activity, constitutes what may conveniently be designated the 'new materialism'. It seems to be the position held by the majority of contemporary men of science.[2]

Yet it lacks the respectability of the mechanical theory. The latter, though untenable, was logically consistent; its denials followed logically from its assumptions. The *new materialism* perpetuates many of the same denials (hence the justice of the designation here proposed) merely by reason of mental inertia, the tendency to continue thinking in old ways whose logical foundation has been destroyed.

Let us clearly recognize that there are only two logical positions in this great issue of purpose and mechanism. You may recognize the validity of both mechanistic and purposive explanations; or you may deny one or other of them. There is no third type of causation or causal explanation in question; no third type has been suggested. The new materialism has no logical standing.

---

[1] *The Order of Nature.*

[2] This position I have discussed at some length as *modern materialism*, the present-day successor to strict or literal materialism, in my *Modern Materialism and Emergent Evolution* (London and New York, 1928). Some confusion arises from the fact that some of the scientists who hold to the new materialism use the word 'mechanistic' as synonymous with 'mechanical'; hence they repudiate what they call the 'mechanistic theory' (meaning the strictly mechanical) while refusing to recognize the causal efficacy of purposive activity. This is not a third position; it is merely a variety of *the new materialism* (as defined above), and might well be called *crypto-materialism*. It is the position of the leaders of the *Gestalt* school of psychology.

RECENT DEVELOPMENTS OF SCIENCE DEMANDING AND JUSTIFYING THE RECOGNITION OF PURPOSIVE CAUSATION

We have seen that reluctance to recognize the causal efficacy of purposive activity has characterized science throughout the modern period; that its chief ground and only logical justification was the belief in the universal adequacy of the Newtonian mechanics to explain all natural phenomena; that with the great renunciation of that belief the reluctance, though now without logical foundation, persists widely among men of science. What then of the empirical grounds for the acceptance of purposive activity? It is naturally in the biological and more especially the human sciences that we may expect to find such grounds. We can hardly expect that the physical sciences shall provide clear empirical evidence of this kind.

It is necessary to be clear about the logical and epistemological principles involved. And here I can only frankly assert the exclusive validity of the pragmatic principle. In interpreting natural phenomena we form hypotheses; when we find that an hypothesis will not work, we must reject it; when we find one that works well, we call it a theory; and the larger the number of groups of natural phenomena for which it provides satisfactory interpretations, the stronger its claim to be regarded as a true theory. There is no other kind of scientific truth, no other meaning to be given to the words.

I pass over the biological sciences other than psychology, merely pointing to the fact that an increasing number of biologists are repudiating mechanical explanations, recognizing the need for explanations of some other type— the organicists, the emergentists, the configurationists, the neutral vitalists, the psycho-vitalists (such as August Pauli, E. Rignano, and E. S. Russell), and the idealists (such as J. S. Haldane).

In this great problem, as in so many others, psychology

is the key science. Psychology during the nineteenth century sought to build itself after the model of the physical sciences with their atomic theories and their exclusively mechanical causation. There is still no one science of psychology, but rather the psychologies of many schools. But the outstanding feature of psychology in the twentieth century is the increasing recognition of the uselessness of mechanical or mechanistic psychologies, and the consequent pragmatic demand for, and increasing influence of, the psychologies which make use of purposive interpretation. We see the rapid spread of the psychologies of Freud, of Jung, of Adler, and of other psycho-analytic sects, all of which are fundamentally purposive (although Professor Freud has needlessly, illogically, groundlessly, proclaimed his adhesion to the deterministic principle). We see the same revulsion against mechanism in the school of *Gestalt*, of which, while some members still hesitate and hedge, others, such as Professor R. H. Wheeler and (less frankly) Dr. Kurt Lewin, plump for purpose. We see it in Professor William Stern's system of Personalism. We see it in the proposal for a *verstehende Psychologie*. We see it in the closely similar demands of German students of the social sciences for a *geisteswissenschaftliche Psychologie* distinct from the mechanistic psychology traditional in the universities. We see it in the latest publications of Professor E. L. Thorndike, long regarded as a great champion of the mechanistic psychology. We see it even in avowed behaviourists, such as Professor E. C. Tolman, who in a recent book elaborates what he strangely calls 'a purposive behaviourism'. We see it in the last book of the late Hugo Münsterberg, published just before his death during the war, in which he recognized that mechanical psychology, though strictly 'scientific', is a perfectly useless and artificially distorted account of human life; a recognition which led him to write two psychologies in one volume, a mechanical in the first half, a purposive in the second.

In short, it is increasingly recognized that the long-sustained efforts of many psychologists to find a mechanistic explanation of the simplest adaptive or intelligent actions of men and of animals have failed and that we can only begin to understand them when we recognize them as instances of striving towards a goal, striving, directed by foresight of some kind and degree and influenced by feeling. The workers in the psychological laboratories are everywhere concerning themselves with problems of motivation, of incentives, goals, drives, determining tendencies, valuations.

I turn to the historians, and here I must be content to present in a few words the substance of an address by Dr. Charles A. Beard which I recently was privileged to hear. Dr. Beard told us how historians, after being dominated by the ideal of a strictly scientific history based on the assumptions of strict determinism and mechanical causation, have at last realized that such history is utterly unsatisfying and unreal, that, if its fundamental assumptions are true, such history is perfectly useless and not worth the trouble of writing it. As he put it, the historians are admitting that they must make history philosophical rather than strictly scientific; he even seemed to suggest that it must become theological once more. All of which is, I venture to suggest, essentially the recognition of the need for a scientific basis of a different kind, a science of human nature, a theory of man, which shall recognize the efficiency of human activities rather than one which represents man as a mere machine, the helpless sport of his environment. In short, the historians, like the psychologists and the workers in the various social sciences, are finding that the mechanical theory of man does not work; that it has led them into a blind alley from which they must return in order to reach the road of progress.

Two pragmatic tests of the mechanical theory of man are in progress on a vast scale. One is the educational system of the United States of America. The other is the attempt of the Russian Soviets to build up a new and better society on this theoretical basis. It is still too early to claim positive evidence from either experiment; but some indications that the theory is not working well are already discernible in both cases.

Finally, I turn to the economists and practical men. Since Adam Smith taught the great principle of *laissez-faire*, that principle has prevailed; and indeed logic, as well as capitalism, was upon that side. For whether the beautiful harmony and automatic self-regulation of the economic system in a purely mechanical world was attributed to the design of the Great Architect or to the felicitous operations of natural selection, the practical consequence was the same. Let man accept the universe as it was, including its economic aspects; for indeed, if he was but one bit of mechanism among the rest, he could do no other. And so, economic man, fortified by the approval of mechanical philosophy, ethics, psychology, and history, gave unrestrained expression to his natural greed and grabbed all he could, under the one great economic law—the Devil take the hindmost. The economist's function was merely to contemplate this part of the universal mechanism, to describe its iron laws and the merciless grinding of its iron wheels; while deprecating those occasional joltings of the economic machine which made a minority of us somewhat critical of its alleged perfections.

And now a bigger jolt than usual has made us all sceptical of the doctrine that man is powerless to direct the course of his own activities, has made us even begin to see that the mysterious economic forces, of which the text-books so long have prated, are nothing other than the desires and aspirations of men. On every hand we hear the demand for economic planning. But, as recently I heard Mr. Norman Thomas insist in an eloquent address, planning is not enough. Before we can rightly plan, we must clearly define our purpose. Planning is but finding a means to a goal. Our common goal must be defined and our common purpose to attain it must be firmly set. Then, if only we can completely throw off the paralysing belief in the mechanical doctrines, we shall go forward to furnish the convincing and final pragmatic demonstration of the causal efficacy of human purpose. Fortunate indeed is the American nation in that it has found and is following a leader who brushes aside all mechanistic sophistry and displays his firm conviction that, through clear and resolute purposing, wise planning, and vigorous action, man, collective man, may make himself master of his destiny.

III
THE APOLLONIAN AND THE DIONYSIAN
THEORIES OF MAN[1]

THE history of speculation on the nature of man is chaotic. The student wanders in a maze of theories propounded by metaphysicians, moralists, theologians, poets, men of science, free-lances, apostles of strange creeds, devotees of weird rites and disciplines. And modern science has added to the confusion by taking seriously the doctrine that man is a machine, and nothing but a machine. Psychology makes claim to be the science which will lead us to the true theory. Yet in spite of much activity directed to the making of such a science, we still have no established theory, no science of man, no psychology, but, rather, many psychologies.

Amidst this diversity of schools, only some knowledge of the history of thought can make the student feel at home and enable him to choose wisely among the many rivals competing for his favour. My aim in this paper is to suggest how the various schools and their historical affiliations may best be understood by relating them to one deep-lying division which we can trace all through the history of European thought; to show that, by following this line of division down the ages to the present day, we can introduce something of order and system into what must otherwise seem a chaotic welter.

The clue I propose to use was first given form by Nietzsche[2] when he pointed to two very different attitudes to life and nature, and especially human nature, expressed in the religious beliefs, the rituals and the arts of Ancient

[1] Reprinted from *The Journal of Philosophy*.
[2] In his early work, *The Birth of Tragedy*.

Greece. On the one hand was that view which is commonly regarded as characteristic of the classical period. It represented man as dwelling in a world wholly intelligible to him, if he would but open his eyes to it and freely use his reason. The world was a fair and smiling scene of which man was lord and master. In this scene man's intellect could subdue and eradicate his passions, base disturbers of godlike reason, and would enable him to guide himself without error or mishap, if only, by its aid, he could find some simple formula which should be the cue to understanding of the physical world. At a very early period Democritus seemed to have provided this clue in his doctrine of the atoms and their perpetual dance. Others sought similar clues; Pythagoras in numbers, Euclid in the science of geometry, Archimedes in mechanics; all these were allied in their aim of achieving a perfectly transparent or intelligible account of physical nature. For thinkers of this type there were agencies of two great kinds, mechanism and reason; and reason could fully grasp and master its only rival, mechanism. It is true there was religion; but in this mode of thinking the gods themselves were fashioned after the image of man; they stood only a little above him through greater knowledge and more complete rationality; and they tended to fade into little more than graceful mythological figures, useful in the arts as symbols and as ideals of human perfection. This way of thinking, the classical Greek way, culminated in the identification by Socrates of virtue with knowledge or the reasonable use of knowledge. It survived the collapse of the Hellenic civilization, and has permeated and in the main has dominated European culture all down the centuries as the tradition of intellectualism. Nietzsche called it appropriately the Apollonian view.

Alongside the Apollonian tradition, orthodox and official, accepted by philosophers and taught in the school and the forum, ran a very different current, one less bright, less clear, less sunny, but one which was perhaps nearer to the truth. In this view, man's reason was but a weak and very fallible part of him, an instrument but little suited to penetrate the mysteries of the physical world, of life and of man's own nature, and still less capable of controlling the dark depths from which his feelings and his actions spring. According to this view, man was not set apart from and above nature, near to the gods by virtue of his godlike reason. Rather he was part of nature; and nature everywhere gave glimpses of powerful forces similar to those he vaguely felt at work within himself, forces tremendous and ruthless, neither mechanical nor rational; forces which might rather be described as blind strivings, insatiable cravings, restless urges towards goals unpredictable, ill-defined and indefinable—forces at once destructive and creative, forces with which man's reason was destined to struggle during long ages before it could begin to achieve control of them by way of some dim and imperfect understanding. For this view, whatever god or gods there be dwelt not apart upon Olympus, careless of the world. Rather the divine creative power that had brought forth all things was immanent in nature, and was an awful mystery, not easily to be understood in anthropomorphic terms, not to be propitiated by gifts and rites, not surely beneficent to man, but inscrutable and dark, ground for fear, though perhaps also for hope. But, though the intellect availed little in face of this dark mystery, some intuitive understanding of it might be achieved by way of intense feeling and emotion sympathetically shared. Hence, alongside the officially accepted religion of Olympus, was another religion, the nature-worship which culminated in the Dionysiac mysteries; and behind Olympus lurked the dim and awful figure of fate or destiny to which the gods themselves were subject. This world-view, contrasting so sharply with the Apollonian, Nietzsche proposed to called the Dionysian.

The Dionysian view of man may be said to have been given sober and scientific formulation by Aristotle, the great biologist, in his essentially teleological biology and purposive psychology, in his recognition, as the fundamental biological fact, of *hormé*, the urge within every creature towards the realization of its specific form and destiny.

To the mind schooled in the science of the nineteenth century it would have seemed fantastic to assert that these two opposed world-views have never ceased to struggle for supremacy and still survive as active rivals. Yet such is the fact; and at the present time these rivals are entering a phase of their long struggle more acute, more defined, more decisive than ever before.

Let us skip the Middle Ages to consider the modern age alone. For the Middle Ages of Europe were a period in which the natural course of development of European thought was distorted and overwhelmed by the dominance of the Christian Church with its authoritarian claim to impose a supernaturally revealed religion.

The Renaissance was a rebellion against the authority of the Church and a revival of the Apollonian world-view. With it came the beginnings of modern science in the work of Kepler and Galileo. Descartes gave to the movement the shape that it was to retain throughout the modern period. He extended to the organic sphere, to the world of life, the strictly mechanical principles of the astronomers. He represented all animal bodies as mechanisms in the strictest sense, as machines and nothing more. By aid of his brilliant foreshadowing of the principle of reflex action, he made plausible the view that every animal action was the outcome of a strictly determined sequence of purely mechanical events. Thus all of nature, every natural event, was made to seem perfectly intelligible. Man alone was an exception; to his bodily machine, extended in the three dimensions of space, was attached an inextended non-spatial being, the rational soul which could interfere with the mechanical processes of the body, guiding them according to the light of reason.

Descartes thus founded the mechanistic physiology, which, beginning with the muscles and the organs of digestion, circulation and respiration, seeking mechanical explanations of all their processes in terms of an atomic chemistry modelled upon astronomy, ascended to the brain and claimed to find all its processes to be equally explicable in terms of the same mechanical principles.

The brilliant successes of the principles of strict mechanism in the physical sciences and their initial successes in biology gave to these principles an immense prestige, which prevailed throughout the eighteenth and nineteenth centuries and only quite recently has begun to wane.

Newton gave precision and new confidence to the Apollonian world-view by showing how the strictly mechanical principles render fully intelligible the motions of the heavenly bodies. The whole of nature was viewed as a vast but perfectly intelligible mechanism, itself created and set in motion once for all by the Supreme Reason. And man, with his mysterious participation in the life of reason, was but a helpless spectator of the strictly determined course of events.

Thus the Renaissance and the Cartesian philosophy led to an age in which the Apollonian view seemed to have finally triumphed, the age of Spinoza's pantheism, a clean-cut piece of Apollonian intellectualism; the age of reason, of the Deists, of Voltaire and Newton and Pope and Godwin.

The guesses of Democritus, the poetic fancies of Lucretius, had become the accepted creed of a science growing every day in authority and power. And it became the task of philosophy, a task with which it has vainly struggled throughout the modern period, to reconcile this scientific view of the world with moral and religious beliefs, beliefs that seemed to be indispensable if man was not merely to live but to live with some hope of living well.

In the eighteenth century, then, the Apollonian view

was triumphant, and psychology, the theory of the nature of man, conformed to the prevailing intellectualism. The cosmos was a vast mechanism; the mind of man merely reflected this mechanical cosmos in an imperfect manner; and the problem of psychology was to give some account of the way in which this passive reflecting took place. John Locke, taking up Plato's great word 'idea', gave it a new meaning by taking over also the ancient theory of perception propounded by Democritus. Ideas are subtle little copies of things which somehow get into the mind through the sense-organs. And, having thus mysteriously entered the mind, there they are, the content of the mind; the mind itself being little more than the empty receptacle, the clean waxen tablet in or on which these copies are received and stored. Then came Hume with his doctrine that there is but one law and one principle according to which the play of ideas takes place, the principle of association. And David Hartley, the physician, presently showed that this principle of association might be regarded as a purely mechanical law of the brain-processes. Thus the reason of Descartes was resolved into the play of mechanical processes, and his soul shown to be a fanciful superfluity.

These simple mechanical principles were carried over to the next century and were actively developed by Bentham, by the Mills, father and son, by Herbert Spencer, and by Bain. In Germany, Herbart gave currency to a very similar psychology, of which also the essence was the mechanical interplay of 'ideas'.

Thus reason, which Descartes and Newton had allowed to man as his sole distinguishing mark, was itself reduced to the play of mechanism, and the Apollonian view of the complete intelligibility of man and nature was apparently perfected. The achievement seemed great; but its cost was greater: for it was made at the cost of destroying reason itself, the absolute competence of which to the interpretation of the world had been its initial postulate.

Man had become a mechanical microcosmos that merely reflects the equally mechanical macrocosmos; his reason an irrational excrescence upon a mechanistic universe, a universe of mechanical events of wonderful complexity, but empty of creative activities.

In the later nineteenth century the Darwinian theory gave new confidence to those who accepted the mechanical interpretation of life, of man, and of reason itself; and psychology became experimental and 'strictly scientific', that is to say, it was modelled upon the physical sciences. It is true that the relation of the 'ideas' (which were said to be the substance of the flux of consciousness) to the brain-processes which they seemed so faithfully to accompany, remained utterly mysterious. But this mystery was for the time successfully shelved by the doctrine of parallelism, or the allied doctrine of epiphenomenalism, according to which the 'ideas' are, as it were, quasi-real shadows cast by the mechanical events of the brain.

Wundt, the founder of the experimental method in psychology, though he claimed to be a voluntarist, contributed mightily to the predominance of the mechanical view of man, largely by rendering orthodox the theory of psycho-physical parallelism. We see this most clearly if we consider the psychology of the late Professor Titchener, of Cornell, Wundt's most faithful disciple, who carried his principles to their logical conclusion. In Titchener's hands psychology becomes, not the science of mind or of human nature, but merely the attempt to render a complete analytic description of states of consciousness conceived as complex conjunctions of ultimate atoms of consciousness; a psychology of processes but no activities; a psychology that refuses to take any account of meanings and motives, and thus, since it has no bearing on the problems of human life, which are essentially problems of meaning and motive, resigns itself, with a sigh of relief, to a position of academic seclusion.

In France the development of psychology ran a parallel

course. French science and philosophy have been predominantly Apollonian, priding themselves on their clarity and rationality. Lamettrie and D'Holbach, the outspoken physiological mechanists; Laplace, with his claim to make all events strictly predictable by the mathematician; Condillac, who converts his mechanical model of a man into a real man by adding to him one sensation after another—these are the typical French thinkers. And we see the tradition culminating in the atomistic mechanical psychology of the earlier works of Professor Pierre Janet, the contemporary dean of French psychologists.

The predominant philosophies of the modern period have run on two parallel courses, both Apollonian, typified by the positivist and mechanist naturalism of Comte and Spencer and the pure intellectualism of the idealists. Consider the culmination of the latter with the English Hegelians. Bosanquet accepted psycho-physical parallelism and the validity of the mechanistic interpretation of life and mind. For him the universe was a mechanical system empty of creative potency and true becoming, redeemed from sheer materialism only by being dubbed 'The Absolute'. Bradley, his colleague and leader, declared that the use of the term 'activity' in psychology was a scandal.

I said just now that Titchener's psychology was the logical outcome of the Apollonian tradition. But certain American psychologists claim the distinction of having carried the process one step farther, a step which, whether logical or not, was natural enough. I mean, of course, the development of behaviourism, a caricature of the Apollonian ideal of perfect intelligibility. All human action is made to seem perfectly intelligible by reducing it to nothing more than a series of mechanical reactions to physical impressions, reflexes and conditioned reflexes; and the stream of consciousness, since it is denied all influence upon these processes, is ignored, as unworthy the attention of any serious person. This ideal result is achieved by resolutely turning the eyes away from all the many facts that will not fit into this too simple scheme, and by refusing to recognize the problems which those many facts so urgently press upon us. This culminating step, I say, was natural enough, so natural that it has found high favour with a multitude of college students, who find it a wonderful solvent of theoretical difficulties and find also that it relieves them, so far as they can really accept it, of all personal responsibility.

So much for the Apollonian tradition and its influence on the theory of man. It has arrived at an impasse; has provoked its own nemesis in the form of radical behaviourism. It has achieved its own *reductio ad absurdum.*

Let us go back to pick up the traces of the Dionysian tradition. At the beginning of our era it was represented by the religion of Mithra, at that time a serious rival of the Christian religion. Christianity itself with its doctrine of sin and redemption may be said to have embodied it. But we jump again to the modern period. There for long it was literally snowed under by the new-blooming of the Apollonian view, which, in spite of its proud claim to sweetness and light, to unimpassioned rationality, was in fact an orgy of intellectualism, whose devotees were intoxicated by the successes of the mechanical theory of the world.

We find only sporadic expressions of the Dionysian view. Jacob Boehme's mysticism was one such. Another is Pascal's famous aphorism: 'The heart has its reasons of which reason knows nothing.' It is chiefly the poets rather than the professed thinkers who express it. It would be easy to cite many passages from Shakespeare in illustration: 'After life's fitful fever he sleeps well'; or 'Sir, in my heart there was a kind of fighting, that would not let me sleep. . . . Our indiscretion sometimes serves us well; when our deep plots do fall; and that should learn us there's a divinity that shapes our ends, rough-hew them how we will.' But the Apollonian current was too strong even for the poets; and Pope's *Essay on Man* and Herrick's trivialities take the place of true poetry, which is borne only on the Dionysian stream. In science it made a feeble and little regarded stand against rationalism in the doctrine of the vital force.

At the end of the century a Dionysian revival found expression in the romantic movement of which Goethe, Wordsworth, Coleridge, and other nature poets were the mouthpieces.[1]

About the same time it flourished in the Scottish schools of philosophy, in the formal works of Ferguson, Hutcheson, Dugald Stewart and others; until Bain capitulated to the rationalism of the English Association school and the Scottish metaphysicians went over to Hegelian rationalism. In France appeared one powerful, but isolated, exponent, Maine de Biran. In England ultilitarianism was dominant. It claimed to be a purely rationalistic philosophy based on the intellectualistic psychology of association. Yet its most powerful exponent, Jeremy Bentham, introduced into it psychological hedonism, the doctrine of pleasure and pain as the springs of all human action, a perverted Dionysian element which never could be intelligibly combined with the pure rationalism of the associaton psychology. In Germany there was a brief outburst of Dionysian thinking in the nature philosophers. In the persons of Oken and Schelling it ran amok, brought disrepute upon itself, and was utterly repudiated by all the academic thinkers.

In the middle of the century came the beginnings of the present Dionysian revival in philosophy and psychology. Schopenhauer, with his teaching of the primacy of will,

[1] Wordsworth is especially interesting in this connexion. In his *Lost Leader*, Mr. H. I. Fausset shows how in the mind of Wordsworth the two types of thinking were in perpetual conflict. 'To discover . . . how the rational and the instinctive may be creatively harmonized is, it seems to me, the most fundamental problem which faces us to-day. And since the cardinal importance of Wordsworth is that he tried and failed to solve this problem as a man, although he came near solving it in moments of divination as a poet, he has, I believe, a very vital meaning for us to-day.'

and von Hartmann, with his mysterious unconscious and its all-powerful impulses, broke away from rationalism and founded a truly voluntaristic or hormic psychology. Lotze, in his restrained and sober fashion, shows something of this influence. And indeed from this time many of the psychologists began, like Wundt, to render lip-service to the Dionysian view—to admit, if only in words, that will and impulse and feeling and emotion are not merely incidental accompaniments or attributes of, nor yet complexes of, sensations and ideas, but rather the very core and foundation of man's being, to which his intellect and reason are accessory and subservient developments. An intermediate unstable position was found by those who, like Brentano and Stout and Külpe, could not content themselves with a purely passive flow of ideas, and, without adequately recognizing the intellect's vast foundation of subconscious impulse and desire, sought some place for true activity within the intellectual processes themselves.

The so-called functional school of American psychologists, represented by Angell and Woodworth, belongs here, presenting as it does a mechanistic basis with some uncertain and incompatible admixture of hedonism and a touch of the truly Dionysian hormic psychology.

It was Nietzsche who, more forcibly than any other, pointed the way to a thoroughly Dionysian view of man, and to a hormic psychology which sees the creative urge to activity as the common foundation of human and animal nature; a psychology which frankly admits the obscurity of this foundation, without seeking to disguise it in a cloud of words, which refuses to pretend to the illusory clarity that comes from the acceptance of rationalism and mechanism; a psychology which recognizes as obscure many problems that truly are obscure; which admits that we are very far from an adequate understanding of man or nature; and insists chiefly that we shall not distort and falsify the immediate teaching of experience in the interests of the spurious clarity and symmetry of a rationalistic system. It gives the name 'instinct' to these obscure racial foundations of our active nature, without pretending that in naming them it has made them intelligible.

Instinct is thus restored from the debased and false conception of it by the rationalists, from Descartes onward, as a mere link in a mechanical sequence, and becomes at once the deepest foundation of human nature and the prime problem of psychology. Man is conceived as having his roots deep in nature and in common with the animals; and the animals are no longer mere machines, but, in their various grades, express, however humbly, the same obscure but powerful forces which have shaped all life, formative, constructive, energic, creative forces, of which we obtain some understanding only through experience of their working within ourselves and by sympathetic intuition of them in other men and other forms of life.

This is the essence of the Dionysian view of man which just now is coming rapidly to the front, displacing the threadbare and sterile view of man as a machine plus reason, or as a machine whose reason is but part of the mechanism. In France, Bergson has given it, in very independent fashion, a powerful advocacy. In the German world we see it attaining great influence in the somewhat distorted one-sided psychologies of Freud and Adler, and in the more rounded and comprehensive psychology of C. G. Jung; for all of whom reason is but a feeble shoot springing from the deep, dim, massive foundation of subconscious strivings of our instinctive nature. We see it in the personalistic psychology of William Stern of Hamburg; in the now very influential school of *Geistes-wissenschaftliche* psychology, of which Spranger is the best-known exponent; also in the *verstehende* psychology of Erisman and Jaspers and others, who frankly renounce the attempt to explain the course of mental life, seeking rather to understand it sympathetically and intuitively.

These schools have rejected the mechanistic intellectualistic psychology, because they have found it useless as a basis for the various social sciences and for application to the practical problems of education and psychiatry. We see a cautious trending in the same direction of the now so influential *Gestalt* or configuration school of Wertheimer and Köhler. And Nietzsche's view is being directly expounded and developed, by Ludwig Klages and Hans Prinzhorn, as a psychology of personality; a characterology, as they prefer to call it, in which intellect (*Geist*) figures as a power hostile and dangerous to instinct and to life itself. Common to all these schools is the repudiation of the mechanistic principles of explanation, and of the attempt to describe our experience as a composite structure of atomic elements, and the recognition of purposive activity, as something other than, and at least as fundamental as, mechanical process. Coinciding in time with this revival of the Dionysian view of man, comes the revolution in physical science, which is showing the inadequacy of the mechanistic principles even in the physical sciences and throwing grave doubt upon the validity, even in the inorganic sphere, of rationalism's favourite dogma, the strict determination of all events.

The physiologists are coming into line. J. S. Haldane, distinguished among British physiologists for his many exact measurements of bodily processes, declares in his recent Gifford Lectures that to say as so many do, in parrot-like repetition, that physiology is revealing the mechanism of life is mere claptrap. In America three at least of the leaders in biology, Jennings and Herrick and Lillie, are in open rebellion against the mechanistic physiology. Two of the most influential psychologists of America are peculiarly interesting in this connexion. William James represented very clearly in his great work, the *Principles of Psychology*, both types, without attempting to reconcile them: the Apollonian in his physiological speculations and his sensationism, the Dionysian in his recognition of the subconscious activities, of freedom of will, of creativity, and in his life-long interest in the varieties of religious experience and in the obscure phenomena of 'psychic research'. In this duplicity of outlook he not only expressed the width of his sympathies, but also he well represents the time in which he flourished.

Münsterberg represented the transition from the one type of psychology to the other in a more explicit manner. Much more concerned than James ever was to attain formal consistency of doctrine, he began by accepting and teaching very confidently the rationalistic mechanical psychology. Then, in middle life, he became interested in the problem of values and in the various practical applications of psychology. Now, in a purely mechanistic world there can be no valuation and no values; and for pure rationalism there is only one value, namely, logical consistency. And, when we come to make practical applications of psychology, we soon find that man is anything but a piece of mechanism which, in some utterly mysterious and passive fashion, mirrors the world about him. Therefore these two interests naturally led Münsterberg to realize the limitations of his mechanical rationalistic psychology and to attempt in his last book to construct a psychology of a radically different kind, namely, a purposive psychology. He went over to the Dionysian tradition; for the clearest, sharpest distinction between the Apollonian and the Dionysian views is that the former leads inevitably to a psychology fashioned after the pattern of the mechanistic atomism now obsolete in the physical sciences; while the latter recognizes the all-importance of purposive striving and traces the development of human volition in the growth of foresight—foresight whose increasing range progressively transforms the subconscious cravings and purblind urges of the animal plane into farsighted intelligent efforts towards distant goals and high ideals.

I suggest, then, that we may most profitably group the

4

many schools of psychology of the present and the past according as they reveal the predominance of Apollonian intellectualism or of Dionysian intuition; finding in this a principle of classification more significant than any other; and especially we find in it a clue with which to thread our way among the many existing rival schools towards the psychology of the future, which, after long wanderings in the wilderness without the pale of science, shall assert its true place as the crown of the biological sciences and the indispensable foundation of all the social sciences.

IV
THE NEED FOR PSYCHICAL RESEARCH[1]

I HAVE the honour to address you in the capacity of President of the American Society of Psychical Research. I have no message to bring you assuring you of the continued existence of the friends you have lost, or of your own survival after death of the body. If I had any such message to announce, I should, I suppose, fill Symphony Hall and charge each of you five dollars for the privilege of hearing me speak. I merely seize this opportunity to put before you the grounds on which, as it seems to me, the American Society for Psychical Research may fairly hope for and justly demand an increased measure of support from the educated people of this country.

I shall not delay to define what we mean by psychical research, nor to sketch the history of the profoundly interesting movement of thought which goes by that name. I will plunge at once into my topic, and will say briefly, first, why we should support psychical research in general, secondly, why this support should be given in particular to the work of the American Society for Psychical Research. Your presence here seems to show that you agree with me in thinking that psychical research is in some manner and degree interesting. My aim is to stimulate that interest, to make clear the grounds which justify it, and to try to give your interest a more practical bent than it has had hitherto.

[1] An address delivered in Boston, May 25th 1922. What was said in support of the American Society for Psychical Research, I would now urge on behalf of the Boston Society for Psychic Research, founded shortly after the delivery of this address and now the one society in America which maintains high critical standards. The address is reprinted from *The Harvard Graduates' Magazine* with the kind permission of the Editor.

The extreme differences of opinion and attitude among us toward the phenomena with which psychical research is concerned make it necessary to emphasize differently the arguments directed to the holders of these divergent views. We may divide intelligent persons into three main groups in this respect.

There are those who refuse to support psychical research because they claim to know that there is 'nothing in it', nothing to be discovered by it. Unfortunately, a considerable number of scientific men, among whom we might fairly expect to find at least the support of sympathy and approval, if not of co-operation, belong to this group. It might seem as though no argument in favour of psychical research could logically touch these persons unless their dogmatic negation can first be shaken. That, however, is not the case. This position of scientific indifference or hostility can be easily turned.

The function of the man of science is, not merely to discover truth for himself, but also to make truth prevail, to establish it in the body of traditional beliefs by which civilized society lives and by which alone it can hope to progress to a better state of things than it has yet attained. And science cannot achieve this great purpose merely by adding one fact after another to the body of scientific truth. It must also examine critically any beliefs which are widely entertained by cultivated minds and by the popular mind and which are, or may seem to be, incompatible with scientific truth; and if on investigation these beliefs prove to be ill-founded, science must give its authoritative verdict against them, and do what it can to overthrow them.

Any man of science who does not admit this to be a proper function and duty of organized science is not worthy of the name; he is merely a man who grubs in a laboratory for his own private and selfish reasons.

The men of science who are opposed to, or indifferent to, psychical research because they profess to know that there is 'nothing in it' beyond illusion and delusion based on fraud, these men really stand upon their belief that the materialistic conception of the world is true. Only from belief in the literal truth of this view can their opinion of the futility of all psychical research be derived; only by that belief can their opinion be justified, in the present state of knowledge. Even, then, if any man of science is convinced of the essential truth of materialism, he is yet under obligation to approve and to give at least moral support to psychical research. For only by a well-organized and long-sustained course of scientific investigation into the phenomena of psychical research can it be proved that there is 'nothing in them'. That investigation is still only in an early stage; and, so far as it has gone, it certainly cannot be claimed to have yielded support to the materialistic philosophy.

If materialism is true, let us ascertain the fact by all means; let the truth be told, though the heavens fall and all the gods also. And let us then hope that civilization may succeed in adjusting itself to this truth and, by its aid, may render human life better worth living. But at present it is clear that the civilized world is becoming more and more acutely divided on this question, the question of the truth of materialism. This lack of sure knowledge, and the consequent wide and widening divergence of opinion is a scandal, a reproach to our boasted scientific culture, and an actual and increasing social danger. Here, then, is one good reason why the convinced scientific materialist should support psychical research.

But there is a second good reason. It is the investigation of the obscure and mysterious and unaccountable phenomena that leads on to great scientific discoveries. Psychical research has already established phenomena which, if they are eventually to be brought within the boundaries of the materialistic scheme of things, will necessarily require and will lead to great developments of that scheme, certainly in the biological, and almost as certainly in the physical, sciences. For this reason also, no matter how convinced he may be of the truth of materialism, the man of science should support psychical research.

For these two good reasons, then, even the scientific materialists and those philosophers who camouflage their acceptance of the materialistic scheme by giving to things names other than those used by science—even they should give their approval and support to psychical research. And, if I were a convinced materialist, I should feel that there was no anomaly in my standing here as president of the American Society for Psychical Research to urge your more active support of its work. I should feel rather that I was merely undertaking an obligation that rests upon men of science.

A second class of cultivated persons, and this is probably the largest class, professes to have no conviction as to the possible results of psychical research. Persons of this class are not prepared to swear that there is 'nothing in it'. Many of them are even inclined to believe that there is 'something in it', and that psychical research may succeed in scientifically establishing the reality of certain forms of existence and happening which science at present officially ignores. But they remain indifferent, prepared to enjoy a good ghost story, or to listen with interest to what may seem *prima facie* a story of telepathic communication. Yet they refrain from supporting, or refuse to support, psychical research. Their attitude seems to be in the main: Why should I dabble in these things? I prefer to leave all such inquiry to those who have nothing better to do.

Now, many persons of this class are not without interest in, or even zeal for, the moral welfare of mankind; and many of them are religiously minded, and perhaps professed Christians. How, then, do they justify their indifference when psychical research says, here is the way to establish the truth, or at least the possibility of the truth, of those beliefs in the reality of spirit and of moral purpose in the world order, on which our moral tradition and moral culture, in fact the whole of our civilization, are founded? Such persons offer two answers. Some say? The reality of spirit, the truth that the world is in some sense a moral universe, the guarantee of the conservation of moral values —these things have been supernaturally revealed once for all, and no further evidence is necessary.

To these our answer is that the evidence of revelation no longer suffices. It may suffice for you individually. But the world at large, especially our Western civilization, is unmistakably drifting away from these beliefs. More and more clearly and with increasing rapidity the purely materialistic view of the world is gaining acceptance, destroying the old beliefs.[1] And psychical research, empirical inquiry into the contemporary evidences of modes of action and being that fall outside the materialistic scheme, such inquiry offers the only possible method of arresting this landslide, of establishing firmly once more in the hearts of men these essential beliefs, the beliefs by which the development of our moral culture has been moulded.

A different attitude is expressed by some of this open-minded but indifferent group. They admit that materialism is spreading and that it is rapidly becoming the real working creed of the mass of civilized men. Yet they say: what of it? Professor Kirsopp Lake gave eloquent expression to this attitude in his recent Ingersoll Lecture. And his expression is the more interesting in that he is a churchman and represents, I suppose, that advanced wing of religious heterodoxy which, as experience shows, becomes the orthodoxy of the quickly following generation. Professor

[1] This, I submit, remains true in 1933, even though two or three physicists have written successful popular books in which they recant some of the dogmatic negations of nineteenth-century physics and express their personal predilection for some vaguely idealistic cosmogony. I would add that throughout this address I use the word 'materialism' in the broad sense defined in my volume *Modern Materialism and Emergent Evolution*, i.e. as covering all purely mechanistic cosmogonies, no matter how immaterial their fundamental postulates.

Lake said, 'I look around and I see men who are essentially materialists leading good and wholesome lives, doing their part as good citizens. They have ceased to trouble about the salvation of their souls and are concerned merely to play the game and to live up to the moral standards they have been taught to accept.' And, he added, perhaps they are all the better for this change of attitude. Coming from a leader of religious thought, this is interesting confirmation of what I said just now of the general spread of materialism.

It was the more interesting because Professor Lake indicated that he himself shared in the general change of belief and attitude, especially as regards the belief in life after death. He indicated that in his view such life has become, in the light of modern knowledge, very improbable; though he added that, if psychical research can produce convincing evidence of its reality, his mind will be open to receive this truth, even though it might demand of him considerable intellectual readjustment. I agree entirely with Professor Lake in his diagnosis of the present tendency; but I am very sceptical in regard to his prognosis. For Professor Lake implied that, in his view, the decent standard of conduct maintained by so many of our fellow-men, in spite of their materialistic outlook, affords ground for believing that civilization and morals may continue to thrive indefinitely on the basis of pure materialism.

That seems to me a very questionable inference. Our civilization, our moral ideals and standards of conduct, have been built up on the basis and under the guidance of certain definite beliefs that are incompatible with materialism, the belief that our lives have a significance and value that is greater than appears on the surface of things, the belief that we are members of an order of things that somehow is a moral order, and that the value of moral idealism and moral effort cannot be measured in terms of material comfort or the satisfactions of our animal nature.

Our moral tradition is the product of such beliefs. There is no good reason to think that, in the absence of such beliefs, any high moral tradition could have been evolved by any branch of the human race. Are we then justified in assuming that, if the foundations are sapped away, the superstructure of moral tradition will continue to stand unshaken and unimpaired, powerful to govern human conduct through the long ages to come?

I gravely doubt it. Any society which continued to prosper in that condition would be living on its capital, its capital of moral tradition. And it seems but too probable that that capital, unrenewed and unsustained by any beliefs other than those permitted by a strict materialism, must undergo a gradual, or perhaps a rapid, attrition.

It is possible even now to point to one way in which the sapping of these beliefs has already seriously modified the moral tradition and influenced the conduct of men so as to constitute a very grave threat to the whole of our civilization. Professor Lake's cheerful, prosperous, decent-living materialists may well think it worth while to live up to the standards of honesty and helpfulness traditional in our society. Finding themselves in the world, through no choice of theirs, they wisely make the best of it; and they see that, in order to make the best of it, they must accept the moral obligations along with the material benefits conferred by civilization. But they are not sure whether in the long run the game is worth the candle, whether, if they had been given the choice, they would have chosen to take a hand in the game. Their attitude is apt to be something of the kind we may express in the words: 'Let me get through with my life honourably and decently'—then after me the deluge. Perhaps few men or women formulate this attitude in words; but it is expressed unmistakably in one great outstanding fact of our civilization. Though each of us came into this life through no choice of his own, each of us can exert choice in the matter of perpetuating the life he bears. We can follow the course of nature and perpetuate our life in our children; or we can refuse to perpetuate it. We can refrain from exercising our privilege of creating new men and women. And the outstanding fact of the utmost significance and evil omen for our civilization is that thinking men and women are choosing more and more to refrain. This is the sign of the times which more than any other casts a dark shadow on the future. And can it be doubted that the decay of religion, with the spread of materialism, is at the root of this refusal to perpetuate the life we bear? Everywhere in history the two tendencies have appeared in close connexion; and together they have destroyed the great civilizations in which they have grown strong.

The case may be simply stated in this way. If materialism is true, human life, fundamentally and generally speaking, is not worth living; and men and women who believe materialism to be true will not in the long run think themselves justified in creating, in calling to life, new individuals to meet the inevitable pains and sorrows and labours of life and the risks of many things far worse than death. Human life, as we know it, is a tragic and pathetic affair, which can only be redeemed by some belief, or at least some hope, in a larger significance than is compatible with the creed of materialism, no matter in how nobly stoic a form it may be held. The fact cannot be gainsaid, and men and women acknowledge it by their actions.

This, I say, is the most ominous indication that a civilization which resigns itself wholly to materialism lives upon and consumes its moral capital and is incapable of renewing it. Here perhaps I may venture on a word of personal explanation. I have two hobbies—psychical research and eugenics. So far as I know, I am the only person alive to-day who takes an active interest in both of these movements.[1] To most of you perhaps these two

[1] Here I owe profound apologies to my friend Dr. F. C. S. Schiller, who has at least an equal claim to this distinction.

lines of scientific study have seemed entirely distinct and perhaps even opposed in spirit. I hope what I have said may serve to show you that, for my mind at least, these are the two main lines of approach to the most vital issue that confronts our civilization—two lines whose convergence may in the end prevent the utter collapse which now threatens.

The indifference to psychical research of this second group is, then, not justified. Unless psychical research —that is to say inquiry according to the strictest principles of empirical science—can discover facts incompatible with materialism, materialism will continue to spread. No other power can stop it; revealed religion and metaphysical philosophy are equally helpless before the advancing tide. And, if that tide continues to rise and to advance as it is doing now, all the signs point to the view that it will be a destroying tide, that it will sweep away all the hard-won gains of humanity, all the moral traditions built up by the efforts of countless generations for the increase of truth, justice, and charity.

The third group indifferent to the claims of psychical research is made up of persons who have become convinced in one way or another, of the reality of the phenomena which psychical research investigates, especially those who believe they have sufficient evidence of the life after death and of communication between the living and the dead. They are the persons generally classed as 'spiritualists'. These fall into two classes; the first, those who are content merely to draw personal comfort and consolation from their belief. With such persons we are not much concerned. They may be classed with other persons who are concerned only with the salvation of their own souls.

But with the other subdivision of this group we are much concerned. Sir A. Conan Doyle may stand as the perfect type of this class. He is a public-minded man, earnestly concerned to gain general acceptance for what he holds to be the truths of spiritualism. But, instead of supporting psychical research, he is indifferent to it; or rather, he is not merely indifferent, he is actually hostile to it. This is very unfortunate; for in this Sir A. Conan Doyle represents a large number of the very best of the spiritualists. This attitude of impatient hostility on the part of such persons is one of the greatest difficulties in the path of psychical research. For experience shows us that, of all those who enter upon the path of psychical research, a considerable proportion become lost to it, by passing over into this hostile camp. Having become personally convinced of the truth of the main tenets of spiritualism, these persons cease to be interested in research and devote themselves to propaganda. It is only too probable that many of those present in this room are inclining to follow this course, that they are hesitating between psychical research and spiritualist propaganda. How are we to meet this very real difficulty? How may we hope to retain the support and co-operation of the already convinced? We cannot afford to lower our standards of evidence or relax the strictness of our rules of investigation, as these persons would have us do. We must continue to run the risk of estranging them by the rigidity of our scientific principles. We must continue to regard research as of the first importance. Our only hope in respect of these persons is, I think, to convince them that, even from the point of view of their main purpose, namely, the spread of what they hold to be the truth, ours is the better plan. If what they teach *are* truths, further research will establish them more firmly; if they cannot be verified by further research, they are not truths and ought not to be taught. And by mere propaganda, by popular lecturing and writing and discussion, they will never succeed in gaining general acceptance for their views, Sir. A. Conan Doyle and those who share his attitude are attempting a perfectly impossible task. They will never convert the world to their view by the methods they are pursuing.

Organized science has become tremendously powerful.

Philosophy and religion, which in former ages were the official dispensers of truth, have had to bow their proud heads before the triumphant march of the scientific method. We cannot hope to stem this conquering advance. We must be content to adopt and to apply the patient and slow but irresistible methods of science. Science has seemed to many minds to lead more and more definitely to the strictly materialistic view of the world. But if that, as many of us believe, is a mistake, if materialism is not the whole truth and the last word of science, only the further progress of science can make this clear to all. Only by the methods of science can we hope to combat effectively the errors of science.

Therefore we confidently say to those who are personally convinced of some or all of the tenets of spiritualism, 'Do not desert psychical research; stand by us, give us at least your moral support. Do not be impatient with our slow methods. Do not be offended by what seems to you our excess of caution, our obstinate scepticism. For our road is the only sure road.' Fortunately we have a few striking examples of men who, although they themselves have attained to personal conviction, have continued unabated their active support of psychical research, such men as Richard Hodgson, Sir Oliver Lodge, Sir William Barrett. Theirs is the example we ask our spiritualist friends to follow.

Now a few words on the reasons why you should support not only psychical reasearch in general, but the American Society for Psychical Research in particular. In this field of research, even more than in any other branch of science, organized co-operation is necessary. Psychical research has to make head against the cold indifference and the open hostility of those who should be its friends, among the men of science on the one hand, and the spiritualists on the other. Therefore 'psychic researchers' need to stand together for mutual moral support. But that is not the only or, perhaps, the chief reason for working together as an organized group. Two or three persons may get together for research, and by the strictest methods they may obtain the most convincing evidence of 'psychic' happenings. Then they may publish a book or a magazine article reporting their observations. A few hundred persons will read it, mostly persons already convinced that such things do happen. And then the whole thing is quickly forgotten. The evidence is practically lost, so far as science and public opinion is concerned.

It is only by bringing together all our evidence in one place, by submitting it to expert criticism, and by inviting the co-operation and corroboration of impartial experts, that any evidence, no matter how good its quality, can be given due weight and added to the growing mass of effective evidence. The American Society for Psychical Research exists to render possible, to facilitate in every way, such accumulation of evidence. And it is, though not perhaps the only society in this country having these aims, the chief of such societies. It offers the best guarantee of effective publication, effective criticism and co-operation. It has a small staff of able and zealous officers; and recently it has secured the co-operation and support, in an advisory capacity, of a council which includes men of great eminence in science and letters. I think you may feel sure that any support you may give to the society, whether moral or material, will be in no danger of being wasted.

And there is yet another good reason why any one interested in psychical research should co-operate with the society. I am expected to say a few words about the dangers of psychical research. I will say only this. An active interest in psychical research is not without its dangers for those who make research in isolation, or in isolated groups of two or three. There is danger under such circumstances of a warping of judgment, of loss of balance, a loss of due sense of proportion. I will venture to assert that you may guard yourselves most effectively against these dangers by working co-operatively as members of the American Society for Psychical Research. In that way you may bring your evidence into an atmosphere of mutual and helpful and cool criticism, which will be a sure safeguard against undue emotionalism and loss of critical balance. Here I may refer to the English Society for Psychical Research. During its existence of more than forty years, it has had the good fortune to have the active co-operation of a number of indefatigable workers. But there is no ground for thinking that any one of these has suffered in any degree any diminution of his intellectual integrity or emotional balance. These workers have been effectively protected against these dangers by the mutual criticism and friendly co-operation they have found within the Society. And the Society has not only protected them against these dangers that beset the isolated worker, but also it has protected them against a risk which all must run who take a hand in psychical research, namely, the risk of imputation of loss of balance. The American Society is capable of doing the same for the 'psychic researchers' of this country. I submit that it deserves your most cordial support.

V

PSYCHICAL RESEARCH AS A UNIVERSITY STUDY[1]

THIS course of lectures on psychical research is, I believe, the first of its kind to be given in any university, whether of this or of any other country; and I venture to think that the innovation will prove to be yet another leaf added to the laurels of Clark University, already so distinguished by its impartial and courageous spirit of research.

Other lecturers, persons distinguished in the most various lines of activity, but all of them qualified by special study of the field of psychical research, will deal with special parts or aspects of the field and from the most diverse points of view. For it is the intention of those who have designed the course that it should represent with perfect impartiality every point of view from which this most difficult and controversial field may be approached; the only stipulation being that each lecturer shall present his facts, his evidences, and his reasoning upon them in a truly critical spirit and with all the impartiality and openness of mind attainable by him.

This course being so great an innovation, it is fitting that this lecture should be devoted to the justification of the inclusion of psychical research among university studies; for there can be no doubt that Clark University,

[1] An address given at Clark University and printed in the volume *The Case for and against Psychic Belief*, Clark University Press, 1927. It may be added that the scheme of this course of lectures was concocted by the late Harry Houdini and myself. Houdini seemed to me to be moved by a sincere desire for the truth and to be fundamentally open-minded. I regard his premature death as a great loss to the cause of psychical research.

while it will be praised by many for its courage and its pioneer spirit in thus opening its doors to a study hitherto denied university recognition, will also be severely criticized by others. It will be said by these adverse critics that the university is encouraging superstition and countenancing charlatanry; that it runs the risk of leading its students into a slough of despair, of entangling them in a quagmire where no sure footing is to be found, where will-o'-the-wisps gleam fitfully on every hand, provoking hopes that are destined to disappointment and emotions that blind us to the dangers of this obscure region; dangers ranging from mere waste of time to disturbance of intellectual balance and loss of critical judgment; dangers which he who enters by the gate we seek to open must inevitably encounter.

Let me begin, then, by frankly admitting that such criticism is not wholly without substance and foundation. The field of psychical research has pitfalls and morasses unknown in other fields of science. The student entering this field cannot avoid contact with vast currents of traditional sentiment, which sentiment, in nearly all cases, he either shares or repudiates with an intensity of feeling that renders calm and critical judgment well nigh impossible. It is as though the student were invited to embark with Coleridge's Ancient Mariner; to exclaim, with him, 'We were the first that ever burst into that Silent Sea'; to witness, with him, strange and even horrible phenomena that seem to defy all the ascertained laws of nature, a phantasmagoria that can have no reality and no origin other than the fantasy of minds disordered by the conflict of strong emotions and blinded by glittering hopes long held before the imagination of mankind, hopes long deferred and now threatened with total extinction by the triumphant progress of scientific inquiry.

Let it be admitted, then, that this is no field for the casual amateur; for the man who merely wishes to take a rapid glance at the phenomena and thereupon form his

5

own conclusions; for the person who approaches it in the hope of finding solace for some personal bereavement; for the dilettante who merely seeks a new and sensational hobby. It is a field of research which at every step demands in the highest degree the scientific spirit and all-round scientific training and knowledge; a field which gives the widest scope for the virtues of the scientific intellect and character and which, just because it makes these demands and affords this scope, is of the greatest value as an intellectual discipline.

Here the mind long disciplined in other branches of science may find the supreme test of its powers and its training, tests of impartial observation, of relevant selection, of sagacious induction and deduction, of resolute discounting of emotional bias and personal influence. Here, better than in any other field, it may learn to recognize its own limitations, limitations of knowledge, of power, of principle; and to recognize also the limitations of science and philosophy themselves, their inadequacy to give final answers to problems which mankind has long answered with ready-made formulae, handed down from the dim dawn of human reflection, and before which men now halt with burning desire for certainty or unsatisfied longing for more light.

The difficulty, the obscurity, the dangers of a field of research are not sufficient grounds for excluding it from our universities. Has not the teaching of all science in our schools and universities been vigorously opposed on just such grounds, on the ground that such teaching might lead young people into intellectual and moral error, or raise in their souls insoluble problems and conflicts that would destroy their peace of mind? That question has been decisively answered. Our Western civilization has definitely repudiated the old way of authority, has committed itself irrevocably to live by knowledge, such knowledge as the methods of science can attain. It cannot return to live by instinct and traditional beliefs;

it has gone so far along the path of knowledge and self-direction in the light of knowledge that it cannot stop or turn back without disaster. The inclusion of psychical research in the scientific studies of our universities is the inevitable last step in this advance from a social state founded on instinct and tradition to one that relies upon knowledge and reason.

But it may be answered by our opponents: the introduction of science to our universities was justified, in spite of its risks, because science offers a mass of well-established truths which are indispensable to the life of the modern State. Psychical research has rightly been excluded because it furnishes no such body of established truth, it has solved no problems, has attained to no sure conclusion.

Let us admit that this contention also is not without substance and force. But to accept it as a sufficient argument would be disastrous. It would imply a false and fatally narrow view of the functions of our universities. It is on just such grounds that the movement against the teaching of evolution takes its stand. It is said that evolutionary biology must not be studied by young people, because evolution is not an established fact, but merely a theory, or a mass of unverified hypotheses. Yet all enlightened opinion rejects this reasoning, rightly holding that the teaching of established truth is only one, and perhaps not the most important, of the functions of a modern university. Such teaching may perhaps be the sole or main function of technical schools. Universities have other, higher, more important functions.

We may, I think, distinguish three main functions of the university, as follows: first, the function of educating the young people within its gates; secondly, the function of research, of extending the bounds of knowledge; thirdly, a function which, as the life of the modern State assumes an accelerating complexity, becomes more and more important, namely, the function of exerting a controlling influence in the formation of public opinion on all vital matters. Consider each of these three great functions in relation to our question: Should psychical research find a place in our universities?

First, then, the educational function. Under this head we may properly distinguish two very different, though inseparable, sub-functions, namely, first, the imparting of knowledge; secondly, intellectual and moral discipline. It is only as regards the former of these that psychical research is open to the indictment of its opponents. Let us admit, for the purpose of the argument, that it has not achieved any conclusions that may be taught as firmly established truths. That admission denies it a role only in what we may roughly estimate as one-sixth of the total field of activity of the modern university, a fraction of the field which is its lowest or least important part.

As regards the other educational functions, intellectual training and moral discipline, it may well be claimed for psychical research that it ranks very high, perhaps highest of all possible subjects of university study. For consider: In what does such discipline consist? First, in attacking problems patiently and resolutely, in spite of failures and disappointments, in spite of uncertainty that any solution may be attainable. Surely, in this respect psychical research may claim a foremost place! No other field of study makes such large demands on the patience and resolution of the student. Secondly, the discipline of observing exactly and recording faithfully phenomena presented to our senses. There is a lower form of such discipline to which the young student of science is extensively subjected, namely, the task of recording as exactly as possible all he can observe within some very limited field; as when he has to weigh exactly some chemical substance, or when he is set down before a microscope and required to draw what is there presented to his view. Psychical research offers little scope for discipline of just this kind; but this is a lower form of observation, one which does not of itself lead to discovery. There is a higher form of observation which requires selective sagacity; it is conducted with a problem in view and under the guidance of some hypothesis which is to be tested. It requires the observer to distinguish the relevant from the irrelevant, to seek out the relevant, to concentrate upon it, and to devise experiments which shall isolate or accentuate it. For discipline in this higher kind of directed observation, psychical research offers unlimited opportunities and makes upon the observer demands of the highest order. Then as regards the reasoning processes by aid of which general conclusions are drawn from the phenomena observed—here the demands upon the thinker in the field of psychical research are very great and the discipline proportionately severe. The physicist or chemist observes the reactions of a single sample of some substance under particular conditions, and is forthwith in a position to state a general conclusion with high probability. The biologist observes some particular feature in fifty or one hundred specimens of some species and, without great risk, makes a generalization which is probably true of all members of the species.

But the psychical researcher is dealing with the most complex and highly individualized of all known objects, namely human beings; before he can summarize his observations in any generalized statement, he must exercise infinite caution, observe unlimited precautions, be ready to allow for an immense range of possible disturbing factors of unknown nature and magnitude. And, when he proceeds to apply statistical treatment to his data of observation, he finds himself facing problems of unrivalled delicacy. For he can never, like other scientists, be content with the comfortable assumption that each of his unit facts is exactly or even approximately equivalent to every other one of the same general order.

If, by reason of the complexity and delicacy of its problems, psychical research rivals all other branches of

science, it far surpasses them all in respect of the demands
it makes on character and, consequently, in respect of
the character-discipline which it affords. It requires
perfectly controlled temper, and a large and understanding
tolerance of human weaknesses of every kind, intellectual
and moral alike; an infinite patience in face of renewed
disappointments; a moral courage which faces not merely
the risk and even the probability of failure, but also the
risk of loss of reputation for judgment, balance, and sanity
itself. And, the most insidious of all dangers, the danger
of emotional bias in favour of one or other solution of
the problem in hand, is apt to be infinitely greater for the
psychical researcher than for the worker in any other
field of science; for, not only is he swayed by strong sentiments within his own breast, but also he knows that both
the scientific world and the general public will react with
strong emotional bias to any conclusion he may announce,
just because such conclusions must have intimate bearing
on the great controversy between science and religion, a
controversy which, in spite of the soothing reassurances
which great scientists and religious leaders now utter in
unison, is still acute and may well become again even
more embittered and violent than it has been in the past.

As regards the second function of the university—the
extension of knowledge—psychical research may boldly
claim its place within the fold; on this ground any opposition to it can only arise from narrow dogmatic ignorance,
that higher kind of ignorance which so often goes with a
wealth of scientific knowledge, the ignorance which permits
a man to lay down dogmatically the boundaries of our
knowledge and to exclaim '*ignorabimus*'. This cry—'we
shall not, cannot know!'—is apt to masquerade as scientific
humility, while, in reality, it expresses an unscientific
arrogance and philosophic incompetence. For the man
who utters it arrogates to himself a knowledge of the
limits of human knowledge and capacity that is wholly
unwarranted and illusory. To cry *ignorabimus* in face

of the problems of psychical research, and to refuse on that ground to support or countenance its labour, is disingenous camouflage: for the assertion that we shall not and cannot know the answers to these problems implies a knowledge which we certainly have not yet attained and which, if in principle it be attainable, lies in the distant future when the methods of psychical research shall have been systematically developed and worked for all they may be worth. The history of science is full of warnings against such dogmatic agnosticism, the agnosticism which does not content itself with the frank and humble avowal that we do not know, but which presumes to assert that we cannot know.

Let us suppose that, after forty years of tentative skirmishing in the wilderness, psychical research, in part as the consequence of this course of lectures, should be received within the scientific fold and systematically cultivated in our universities; and suppose that, after a hundred years of such cultivation, its representatives, surveying the results of all the work done, should find themselves compelled to utter a purely negative verdict, to assert that psychical research had attained to no positive answers to any of the problems it had set out to solve. What then? We still should have to repeat: 'There is the gate to which we have no key; there is the veil through which we may not see.' But, also, we still should have to add: 'And there the Master-knot of Human Fate!' And, though science might then turn aside, baffled and discouraged, it would at least have given some respectable foundation for the cry *ignorabimus* and have made some real contribution to our knowledge of the limitations of human knowledge.

But, some hearer will object, this question of the limits of human knowledge is one not for science but for philosophy; and in all our universities philosophy has long had its numerous representatives and a well recognized place: it is for the philosophers to answer the questions which science leaves unsolved. Such an objection would imply an old-fashioned and quite mistaken view of the scope and functions of philosophy.

Philosophy may rightly claim to teach us how to think, how to live, and how to die. It may answer the question, Given the present state of the world and of our knowledge of it, what ought I to do? But it is wholly incompetent to answer the questions, What may I hope? What may I expect? A cosmology that is to be more than a fanciful speculation must be a scientific cosmology; and, as science progresses, our cosmology must change with it. Every cosmology that professes to be philosophical rather than scientific is a hollow pretence. Only science working by the methods of science can presume to answer the question, What is? Philosophy must learn that its proper field is defined by the question, What ought to be?

And here I will ask leave to revert to the disciplinary, the educational, function of psychical research with special reference to students of philosophy. In my opinion, formed through considerable contact with such students, their chief lack is lack of knowledge of science; and of all forms of science that which can most enlighten them is psychic research. Its literature can go far to induce in them that which so many of them need, namely, a clear recognition of the limitations of the scope of philosophy and a corresponding humility in themselves as philosophers. For here they will find that questions which philosophers through all the ages have answered in their peculiar and utterly diverse fashions and in the main according to their obscure predilections, are capable of being approached by the methods of science; and the mere act of following in imagination such lines of approach can hardly fail to bring home to the student the fact that the methods of philosophy, divorced from science, are of no avail. He will be brought to realize that philosophy, whether it aims to sketch the main features of the universe or seeks to instruct us regarding the values and the duties of mankind, must, in both cases, proceed from the fullest possible knowledge of what science has achieved, or lay itself open to those charges of futility and ignorant presumption which so often have been launched against it.

What, then, are the essential questions on which we may expect new light from psychical research? They may all be resumed in one, namely, Does mind transcend matter? Or more fully stated, Is all that we call mental, intellectual or spiritual activity, is all understanding and reason, all moral effort, volition, and personality, merely the outcome and expression of a higher synthesis of physical structures and processes and, therefore, subject to the same general laws and interpretable by the same general principles as those which physical science arrives at from the study of the inanimate world? Or are mental activities, are all or some of the essential functions of personality, in some degree independent of the physical basis with which they are so intricately interwoven? Have they their own peculiar nature, interpretable only in terms of principles quite other than those whose validity has been proved by the victory of man over his physical environment?

It is the old problem of materialism versus spiritualism or idealism, of mechanism versus vitalism in biology; or, as I would prefer to formulate it, the problem of animism versus mechanistic-monism.[1] This has been the central problem of philosophy for more than two thousand years; and always the philosophers have been pretty equally divided into two groups, those who say 'Yes' and those who say 'No'. The course of development of modern science has on the whole tended strongly to give predominance to the view which denies the transcendence of mind. Idealistic philosophers have struggled in vain to stem this tide, urging that it is absurd to regard mind as subject to the laws formulated for the interpretation of physical

[1] This formulation of the problem is explained and defended in my book, *Body and Mind: A History and Defence of Animism*, London, 1911.

phenomena; for, they say, it is the mind of man which conceives the physical world and which has itself in some degree created those phenomena.

But this and all similar reasoning remains inconclusive and must ever remain so. We are up against a question of empirical fact; and the answer to the question can be brought only by the methods of empirical science.

Many of the greater physicists have inclined to think that their own science points towards a positive answer to this question of transcendence; and it is possible that the progress of physical science and of biology may in the course of time lead us to a decisive answer to this central problem. But, if so, the answer will be achieved only very slowly by very indirect methods of attack. The essence of psychical research is the proposal to attack the problem directly. If mind in any manner and degree transcends the physical world and its laws, surely it may, somehow or somewhere, be possible to obtain direct evidence of the fact by the methods of science, by observation of phenomena and by reasoning from them! That is the proposition on which psychical research is founded. Psychical research proposes, then, to go out to seek such phenomena, namely phenomena pointing directly to the transcendence of mind, and, if possible, to provoke them experimentally. Phenomena of this kind have been reported in every age; and in every age antecedent to our own age, dominated as it is by the principles of scientific evidence, their obvious implication has been accepted. Psychical research proposes to marshal all such sporadically and spontaneously occurring phenomena, to examine them critically, to classify them, to discover if possible the laws of their occurrence and to add to them experimentally induced phenomena of similar types.

Consider now the third great function of our universities, the guidance of public opinion. It is perhaps from this point of view that the admission of psychical research to the universities is most urgently needed. Here is a most obscure question vitally affecting the intellectual outlook and the moral life of men in general. Surely it is for the universities to find, if possible, the light that we need! What ground can be found to justify their neglect or repudiation of the task? Several such grounds are implied, though rarely formulated explicitly.

First it may be said, the task is one for the philosophers and theologians, who are well represented in the universities. But philosophers and theologians have wrestled with it for long ages; and there is no faintest reason to believe that, by their methods alone, they can achieve in the future any greater success than they have attained in the past. Let us glance at the grounds they offer us for accepting a positive answer in face of the general tendency of science to insist on the negative answer. They may all be reduced to two. First, the moral ground; to believe in the transcendence of mind is a moral need of mankind in general. Such belief, it is said, is essential to the maintenance and progress of our civilization. Our civilization has been built up on a foundation of, and under the sway of, such belief; and, if that foundation and that influence should be taken away, our civilization must surely decline; even though it be possible for exceptional individuals to continue to attain high moral excellence in an attitude of stoic agnosticism. This argument is respectable; it has weight and substance. Given a balance of evidence and the impossibility of assured knowledge, we would be justified in accepting that view which seems the more conducive to human welfare. This argument, which perhaps William James was the first to state and defend explicitly, is, I suppose, implied by those who ask us to continue to accept the transcendence of mind as an article of faith. But this moral argument in no sense justifies a refusal to countenance or support psychical research, which is nothing less than an endeavour to replace faith by knowledge in this matter. If, from time to time, religious leaders exhort their flocks to eschew psychical research and pour scorn upon it and all its works, we cannot wholly acquit them of a preference for ignorance over against knowledge. It would seem that they fear the result of psychical research; they fear either a negative outcome of the great inquiry, or a positive outcome which shall disturb the minds of their flocks by bringing knowledge not strictly in accord with traditional beliefs. Therefore they ask us to remain content to accept these beliefs on authority. But it is too late to advocate that policy with any hope of success. As I said before, it is obvious that we have left the age of authority behind and that our civilization is irrevocably committed to the attempt to live by knowledge, rather than by instinct and authority.

Consider now the second main ground offered for acceptance of the positive answer. If we ask—Whence does ecclesiastical authority derive the views it seeks to impose—the answer is that they are founded upon alleged historical events of a remote age, events of just such a nature as psychical research is concerned to investigate at first hand as contemporary events. However we regard the evidence of those remote events, we can hardly claim that the lapse of some two thousand years has made the evidence of them less disputable; and in any case it is clear that mankind in general is ceasing to find that evidence sufficient. More and more we are inclined to say—You ask us to accept the transcendence of mind because we have certain dim records of events which occurred two thousand years ago, and which, if the records be above suspicion, would seem to justify and establish that belief; and yet you would forbid us to examine, in a candid and critical spirit, similar events that are reported as occurring among friends and neighbours. Truly, he who repudiates psychical research in the interests of religion and of religious authority cannot easily be absolved from the charge of a timid obscurantism!

But it is not only in respect of this high problem of

transcendence that public opinion needs from the universities guidance of a kind which they can give only if they cultivate psychical research. That after all is a problem for the intellectual few; although the views of those few may have far-reaching influence upon the lives of the many. The great public does not much concern itself with the question—Are we truly in some degree rational beings capable of moral choice and creative endeavour? In the main they continue to regard themselves as such beings, in spite of all statements of scientists and philosophers to the contrary. But they are much concerned to know what kind and degree of influence mind can exert upon bodily processes, what truth there is in the claims of many sects and schools of mental healers. They do keenly desire to know whether there is a kernel of truth in the widely accepted claims of communication with departed friends; whether each of us, as science tells us, is for ever shut off from all his fellows by the distorting and inadequate means of communication provided by sense-organs and muscular system; whether there is not some common stock of memory and experience upon which men may draw in ways not recognized by science; whether at death each of us is wholly exterminated; whether ghost-stories are founded only on illusion and other forms of error.

There is in all lands an immense amount of eager questioning about such matters; immense amounts of time and energy are given to ineffective efforts to obtain more light on such questions. And unfortunately there is a multitude of persons who for the sake of filthy lucre take advantage of these eager desires, these strong emotional needs, and of the prevailing lack of sure knowledge, to falsify, obscure and fabricate the evidence.

It is perhaps this last aspect of the present situation which most urgently calls for action of the universities. In spite of the immense and growing prestige of science and its steady and scornful negative to all such questioning, the whole civilized world increasingly becomes the scene of a confused welter of amateur investigation, of conflicting opinions, of bitter controversies, of sects and schools and parties, each confidently asserting its own views and scornfully accusing the others of error, and of woeful blindness or wilful deception.

The negations of the scientific world are of little or no effect upon this chaos of conflicting beliefs and ardent desires. And so long as science stands apart, coldly refusing to take a hand in the game, refusing to take seriously the questions asked, refusing to bring to bear upon the many phenomena that keep alive these conflicts, these hopes and these beliefs, its powerful highly organized apparatus of investigation, its negations will continue to exert but little influence towards stilling the tempest and clearing the confusion.

Let me state the demand upon our universities at its simplest and lowest. Let us suppose that we are firmly convinced that no positive knowledge is attainable, that the outcome of a sustained, organized, and co-operative attack upon the problems of psychical research, such as the universities alone are capable of making, must lead to purely negative conclusions; I submit that, nevertheless, we ought to recognize such inquiry as a task which the present state of fervid chaos in the public mind urgently requires of the universities that they undertake and steadfastly pursue.

The situation, its needs and its demands on the universities may be illustrated on a small clear-cut scale by one particular problem which has long been recognized as crucial in psychical research, namely the problem of telepathy. Does telepathy occur? That is to say, Do we, do minds, communicate with one another in any manner and degree otherwise than through sense-organs and through the bodily organs of expression and the physical media which science recognizes?

Science asserts that no such communication occurs or can occur. Yet, in all ages antecedent to our own, belief in such communication has been universal. And in our own sceptical age and community, such belief is still very general. It is held by all intelligent Christians; for it is implied in the practice of prayer and communion. A very large proportion of intelligent educated persons believe they have observed or experienced instances of such communication. In that highly educated, scientific and sceptical class, the medical men, it is, I think, true to say that about one in three believes that he has first-hand knowledge of indisputable instances of it. A careful, highly critical statistical survey of such sporadic instances, made by persons of the highest qualifications, has resulted in a strongly positive verdict. A number of carefully conducted attempts to obtain evidence of it under experimental laboratory conditions have given equally positive results. A number of men of great distinction and of the highest intellectual and moral qualifications have announced themselves as convinced, after due inquiry, of its occurrence. Yet, in spite of all this, science, especially science as represented in the universities, refuses to regard the question of its occurrence as one to be taken seriously, as one deserving of investigation. And why? Simply because we cannot at present see how such communication can take place.

Now, to deny the possibility of events of a certain kind on the ground that we cannot understand how they may be brought about is very unsatisfactory even in the sphere of physical science. It is still more unsatisfactory and positively misleading in the biological sciences. And in relation to any events in which the human mind or personality plays a part, it is reprehensible and utterly inadmissible as a ground of denial or refusal of investigation.

What more suitable task for a research department of a university can be conceived than the task of investigating such a problem. The individual man of science may and does offer two valid excuses for ignoring this and other problems of psychical research. He may say, This is not my line, I have other things to do. Or he may say, I have tried and have had purely negative results. But our universities as a group of national institutions cannot excuse themselves in this way. The signs of the times call aloud to them that they shall follow the courageous lead of Clark University, shall frankly acknowledge their responsibility and welcome psychical research to an honoured place within their gates. Nowhere else may we hope to find the calm critical temper of scientific inquiry sufficiently developed and sustained; to no other institutions or associations can we hopefully entrust the task of shedding the cold clear light of science upon this obscure and much troubled field of vague hopes and vaguer speculations.

In conclusion, greatly daring, I will venture to say a few words in reply to a question which I feel sure many of my readers wish to put to me, the question, namely—In your opinion has psychical research hitherto achieved any positive results? I am not the sort of person who holds a great number of clear-cut positive and negative beliefs. I am rather a person of the kind that deals in probabilities and degrees of probability, recognizing that our best formulations are but relatively true, that human mind and speech are incapable of formulating absolute truths. Therefore I can attempt in all frankness only qualified answers. In my view the evidence for telepathy is very strong; and I foretell with considerable confidence that it will become stronger and stronger the more we investigate and gather and sift the evidence. In my opinion there has been gathered a very weighty mass of evidence indicating that human personality does not always at death wholly cease to be a source of influence upon the living. I am inclined to regard as part of this evidence the occurrence of ghostly apparitions: for it seems to me that, in many of these experiences, there is something involved that we do not at all understand, some causal factor of influence other than disorder within the mental processes of the percipient. I hold that a case has been made out for clairvoyance of such strength that further investigation is imperatively needed; and I would say the same of many of the alleged supernormal physical phenomena of mediumship. I am not convinced of the supernormality of any of these in any instance. But I do feel very strongly that the evidence for them is such that the scientific world is not justified in merely pooh-poohing it, but rather is called upon to seek out and investigate alleged cases with the utmost care and impartiality.

To some of you this confession will seem to make extravagant claims for psychical research; to others it will seem that I am quite sceptical. Such wide differences of view will continue to divide us until the universities shall have brought order, system, and sustained co-operative effort into the domain of psychical research.

POSTSCRIPT WRITTEN IN 1933

I am glad to be able to report that the Duke University, which I have had the honour of serving since 1927, has followed the lead given by Clark University and has gone farther in the same direction. Psychical research has been not only prosecuted in the laboratory with very interesting results, but also furthered by a course of lectures, given by my colleague Dr. J. B. Rhine and eagerly followed by a large class of students. Also the university has granted the Ph.D. degree to one graduate student of the department of psychology who has worked faithfully for many years at a psychical research problem and embodied his work in a thesis submitted in application for the degree.

VI

ANTHROPOLOGY AND HISTORY[1]

IN choosing, as my topic for the Boyle Lecture, 'Anthropology and History', I felt that it was peculiarly appropriate. For the man whose memory is celebrated by us to-day, great as he was in physical science, did not confine his interest and his researches to that sphere; he was keenly and actively interested in the great problems of human welfare; indeed he strictly judged the value of scientific discoveries by the standard of their power to contribute to such welfare. In his day anthropology had hardly begun to take shape; there was no science of man capable of throwing light on the problems of the historian and the statesman. But we may feel sure that, if he were among us now, he would be a foremost leader in that field of thought to which I wish to direct your attention this evening.

With the growth of knowledge there goes an increase of specialization in the pursuit of knowledge. This specialization is reflected in the organization of our university studies. Though inevitable and necessary, such specialization has serious drawbacks which are too obvious and familiar to require to be pointed out by me. The proper antidote for these drawbacks is to bring into prominence the wide background to all our studies, to take a *vue d'ensemble*, a synoptic view, in relation to which our detailed researches and discoveries may fall into their true perspective. They are then seen to be but very small affairs in themselves, but to be nevertheless fragments of the great whole of living advancing knowledge; and it

[1] The Boyle Lecture delivered in Balliol College, Oxford, May 1920.

82

becomes clear how from their relations to this whole they derive their dignity and worth.

In this university it is our prime function to preserve, to foster, to promote this relation of the parts of knowledge to the whole, to rectify the effects of the inevitable increase of specialization of studies, by making their background as wide, as synoptic as possible, keeping it ever before our minds. A notable step in this direction is, I believe, contemplated by the university, namely the institution of a new school of modern history and philosophy: for this conjunction of history with philosophy means that history is to be studied from the broadest human standpoint; not as the history of political institutions, or of economic relations only; but as a department of anthropology, the science of man. I wish in this lecture to try, as a humble student of anthropology, to illustrate the way in which history may be enriched, and the effects of inevitable specialization and increasing concern with minute historical facts may be balanced and corrected, by viewing it on the broad background of anthropology, by treating it as a department of that wider science.

Anthropology is too commonly thought of in a vague way as concerned solely or chiefly with the measurement of skulls and with endless disputes as to the significance of such measurements. That of course is a grotesque notion of the science. But it is very generally conceived, even by some of its professional exponents, in a way which is far too narrow, namely as the science which is concerned only with primitive or savage man. So conceived it has a vast field of research; one which urgently demands to be cultivated, for this reason—that primitive man and primitive societies are rapidly ceasing to exist; they are being exterminated or transformed. Very soon the opportunities which still exist for the observation of the life of men in a state of primitive culture will have passed away. That is the ample justification for giving prominence in anthropology, for giving in fact an almost exclusive attention in our schools of anthropology, to the study of primitive man.

And our excellent school of anthropology, under the wise guidance of Mr. Balfour and Dr. Marett, has shown how stimulating, how attractive, and how productive of new knowledge and understanding this study may be made, when given a due place in our university. But, though the study of primitive man thus rightly figures at the present time as the most prominent part of anthropology, that is a consequence, first, of the urgency of such study (on the grounds just now indicated); secondly, of the present stage of development of the science. For the science is still in an early stage of development, and it is right that it should attack its problems where they are presented in the simplest forms, namely, in the small, relatively isolated, communities of primitive man. For there the races of men are found in relative purity, there institutions are relatively simple, and there the human mind works relatively free from that accumulation of complex tradition which is civilization.

But anthropology is gathering knowledge and understanding of primitive man, not for his sake alone or chiefly, but rather for the sake of cultured man. All the understanding of primitive man that the science is so industriously and brilliantly achieving is of value chiefly because it throws light upon the nature and the institutions of civilized men, of ourselves. This it can do only when brought into working relations with history in the narrower sense of the word. And I venture to think that, without the co-operation of anthropology, history cannot achieve its legitimate triumphs; cannot hope to understand the rise and fall of peoples, states, and nations; cannot forecast the future, or afford to statesmen in full measure the guidance which they seek in the study of the past, when they undertake to shape the course of the future.

History has been slow and cautious in calling anthropology to its aid, and rightly so; for attempts to apply anthropology have been made with disastrous effects. The latest and most disastrous of the attempts may be said to have begun with the publication by Count Gobineau in 1854 of his notorious book, *L'Inégalité des Races humaines*, to have culminated in the still more notorious work of H. S. Chamberlain, *The Foundations of the Nineteenth Century*, and to have played no inconsiderable part in that systematic debauching of the German nation which brought on the Great War. These writers and the school which they represent taught that of all races of mankind one is greatly superior to the rest, that it alone is capable of creating and maintaining a high and noble civilization; that, therefore, it ought to rule over all other races and peoples of the world. Putting aside the question whether there may be a small grain of truth at the bottom of this dogma, which so extravagantly exalts the Nordic race of Europe, we must admit that such rash and ill-balanced attempts to apply anthropology to history have done great harm. For, apart from their disastrous use in perverting the culture of Germany, they have discredited anthropology in the eyes of all sober historians, and in that way have greatly postponed the day when history and anthropology may enter into full and fruitful collaboration.

This and similar rash attempts, based on insufficient data and on conclusions distorted by racial and national prejudice, have brought anthropology into some disrepute and have led many sober thinkers to reject all consideration of the problem of racial peculiarities in their bearing upon history, and to regard all attempts of anthropology to throw light on history as the work of men whom they contemptuously describe as 'race-theorizers'. In reaction from these rash applications many thinkers have gone over to an opposite extreme view: either they deny all differences between races and peoples in respect of mental constitution; or they affirm that the degrees of such differences as obtain are too small to affect the destiny of nations, and too obscure to be discerned or defined by any methods we may hope to devise. J. S. Mill gave expression to this view when he wrote: 'Of all vulgar modes of escaping from the consideration of the effect of social and moral influences on the human mind, the most vulgar is that of attributing the diversities of conduct and character to inherent natural differences.'

This dictum has been widely accepted; and it is, I think, broadly true to say that in recent decades historians have treated the racial composition of peoples, or at least of civilized peoples, as a negligible factor. Mill and those who have followed him in this respect have believed that in civilized societies education and social environment so powerfully mould the development of each generation as completely to override any racial peculiarities, rendering them of no account. Therefore, without dogmatically denying the existence of racial differences, they regard the populations of all civilized states as made up of men and women who, for the purposes of history and sociology, may be regarded as so much indifferent raw material —a mass of units, all made to the same pattern, each one the exact equivalent of every other one—such differences as individuals or classes may show being wholly due to differences of opportunity and nurture.

Yet, after paying some attention to this question, I am firmly persuaded that this opinion is false and that racial composition and the changes that are going on, and always have gone on, in the innate qualities of peoples are factors which are of the first importance for the understanding of history; that they afford the key to some of the most striking events in the history of Europe and of the world. In one short lecture I cannot hope to expound, still less to defend and justify, this view. I can only attempt to indicate some of the main points of the long argument that would be necessary, in the hope of opening your minds to serious consideration of the evidence.

Those who deny all importance to the racial composition

of peoples, and whom perhaps you will allow me to call the 'race-slumpers', in order to distinguish them conveniently from the 'race-theorizers', commonly insist on certain undeniable facts. They point out that any racial differences of mental constitution, if such exist, are so slight and obscure that no one has been able to define them convincingly. And they insist on the closely connected fact that, when a man of any race you please, be his skin white or brown or yellow or black, and his head round, long, or square, is brought up in a civilized community, he commonly acquires all the elements of its culture with complete facility, and becomes indistinguishable in mental traits from the mass of the population amongst which he lives. Further, they are never tired of pointing out that in Europe there are at the present day no pure races, that the population of every European country is made up of descendants of many races and stocks, blended more or less intimately by intermarriage.

These facts must be admitted; but it does not follow from them that racial composition is of no importance. First, let us recognize that recent developments of the science of evolution and heredity are on the side of race qualities. At the time when Mill, Buckle, and others taught that race is a negligible factor, 'genetics' was hardly born. It was assumed that the experience of each generation impresses itself upon the race, moulding it rapidly and effectively to harmony with its environment, physical and moral. We know now that this was an error. We know that racial qualities, both physical and mental, are extremely stable and persistent, and that, if the experience of each generation is in any manner or degree transmitted as modifications of the racial qualities, it is only in very slight degree; so that any such moulding is effected only very slowly in the course of many generations.

We have learnt also the error of another of the fundamental assumptions of writers of the school of Mill—the assumption, which they inherited from Locke — namely, that the mind of each new-born babe is practically a *tabula rasa*, a clean sheet upon which experience may write what it pleases, an indifferent plastic mass which may be moulded into any shape by the pressure of its environment, because it has no well-marked hereditary tendencies or peculiarities that can resist, or react in specific ways upon, this moulding pressure.

I would submit for your consideration a principle which seems to me of the first importance in this difficult field, a principle the neglect of which is, I think, responsible for the continuance of the extreme divergence of views, those of the 'race-theorizers' and those of the 'race-slumpers'. The principle is that, though peculiarities of racial mental qualities are relatively small, so small as to be indistinguishable with certainty in individuals, they are yet of great importance for the life of nations; because they exert throughout many generations a constant bias upon the development of national culture and institutions; a bias which, though its effects may be small within the lifetime of one or two generations, yet in the long run produces very great effects, gradually bringing the culture and institutions of a people into more complete harmony with its racial qualities. The culture of every civilized people is made up of many elements. Each such element either has come from some source external to that people or has been originated by the thought of some one exceptional and original mind, or some few such minds, within the people. In either case such a novel element of culture becomes established in the tradition of a people only through a process of competition with other rival and incompatible elements, which it supplants or prevents from becoming established. And a main factor in determining success and survival in this struggle for life, this process of natural selection among culture-elements, is always the racial composition or, in other words, the sum of the innate mental qualities of the people. For this sum of innate qualities is the environment to which, by way of a subtle and long-continued process of natural selection, the culture-species must perpetually adapt itself.

If this principle be sound, two important consequences may be drawn from it, both of which may, I think, be amply verified by observation. First, it follows that the racial qualities of a people are expressed more clearly and fully in their culture and institutions than in the lives of individuals. Secondly, it follows that the racial peculiarities of any people, being in the main in harmony with their culture and institutions, are developed and accentuated in each generation by all the cultural influences under which it grows to manhood. In consequence of this harmony and common direction of the innate and of the acquired qualities of any people that has undergone a long course of natural development, it becomes extremely difficult to distinguish between them; and it is made easy for the 'race-slumpers' to attribute to the influence of nurture alone those national peculiarities which are in reality the combined expression of both nature and nurture, harmoniously co-operating.

There are rare instances in which an alien culture has been imposed upon a people by external power and authority. And in some of these we have the opportunity to observe and to distinguish clearly the influences of racial qualities and of culture, respectively. Such an instance we may see in Hayti, where the French, in their early enthusiasm for the ideals of liberty, equality, and fraternity, imposed on a population of Negro race much of the culture of Western Europe, including Christianity, the French language and schooling; and then, after a brief period, withdrew their controlling hand and left the people thus generously endowed with an alien culture to work out their own destiny. The results are known to all the world, namely, a rapid relapse into barbarism, with frequent outbreaks of voodooism, cannibalism, and other savage practices, and a political life such as is only to be paralleled by the fanciful efforts of comic opera.

A second very similar and equally instructive instance is to be found in the republic of Liberia.

Let me illustrate the principle I have enunciated by reference to two instances of the spread of religious institutions through large areas of the world. Religion is often regarded on the one hand as a very personal matter, in which each man's own soul seeks its own way of reconciliation and salvation, according to its own peculiar nature. Yet on the other hand we all know that for the vast majority of men the forms of their religious beliefs and practices are determined by those of the society into which they are born. There you have the opposed views of the 'race-theorizers' and of the 'race-slumpers' in their application to a particular problem.

Now the religion of Buddha was founded in India and spread rapidly among the peoples of that sub-continent. But, after this initial success, it declined again in popularity; and it has almost died out in its original home. Among the people of a very different race farther east, namely the peoples of Burma, of Tibet, of China and Japan, it continued to spread; and among these peoples it has maintained itself for many centuries as the dominant religion. It is difficult to suggest any grounds for this fading away in one great area and continued spread and dominance in another, other than the wide racial difference between the populations of these areas.

More striking perhaps, because nearer to us in time and place, is the coincidence of the distribution of two forms of the Christian religion, the Roman and the Protestant forms, with the distribution of the races of Europe. Out of all the researches and disputes of the ethnographers on the races of Europe, one fact emerges most clearly as well established, namely, the identity of the Nordic race, characterized physically by fair complexion, tall stature, and longish skull, and mentally by a high degree of independence and initiative and by love of adventure or restlessness. I am not asking you to follow Gobineau and

Chamberlain and to assign to this race all the virtues, while denying them to the other two great races of Europe, the Alpine and the Mediterranean. The mental qualities I have mentioned are so well marked as to be recognized and admitted by almost all who have given attention to the subject. Well, if you study a map of the distribution of Protestantism in Europe and another of the distribution of the modern representatives of this race in Europe, you will find that the two are almost coincident. Great Britain, Scandinavia, Denmark, Holland, and North Germany, and in lesser degree North-Eastern France, are the principle components of both these areas. I am not asking you to believe that each Protestant chooses his form of religion because his head is long or his eyes blue, nor even because he has the independence and impatience of authority characteristic of the Nordic race. But I say that these racial qualities must be believed to have played an appreciable part in determining the success of the Reformation among those populations and communities in which this racial element was strongly represented.

There are other statistical facts which reveal the persistence of these mental qualities of the Nordic race in the populations in which the physical features of the race preponderate. Of these I will mention only that the restless spirit of adventure is expressed in such populations by high rates of drunkenness, suicide and divorce, rates far higher than in other populations closely similar in all other conditions, but differing in the relative absence of these physical traits.

Consider now the application of this anthropological fact to the explanation of a political phenomenon of the greatest importance and interest. France and Great Britain have been the two greatest colonial powers of modern Europe. Both of them have made vast colonial conquests. But how great is the difference in the subsequent histories of those conquests! The British have established themselves wherever they have set foot, and their colonies have continued to expand up to the extreme limits possible. France on the other hand has lost vast areas which once were hers; and in those which she retains the number of men and women of French blood is comparatively small. Further, at each of the many points where British and French colonists and empire-builders have come into direct rivalry and conflict, the British have succeeded at the cost of their French rivals. It has been the same story in India, in Africa, in America, in the Pacific Ocean. This is not due to any intrinsic lack of courage or genius or ambition on the part of the French. The genius of the French people in all the arts and sciences of civilization stands supreme, or at least as high as our own.

The different powers of expansion throughout the world shown by these two peoples must be correlated with a profound difference of national character. This difference is so striking that it has been universally recognized. Buckle summed it up in two phrases, 'the spirit of independence of the British' and 'the spirit of protection of the French'. The difference of national character or of mental quality of the two peoples, conveniently denoted by these phrases, shows itself in almost every detail of the political institutions and social customs of the two peoples, most notably perhaps in the highly centralized administrative system of France and in the family customs of her people.

I have no time to illustrate the difference in detail; but must pass on to insist that the difference is a deep-lying racial difference and not one merely impressed upon each generation afresh by education and social environment. Many attempts have been made to explain the difference in this way, but they have signally failed. Most of such attempts have consisted in pointing to the highly centralized administrative system of the French and to their various social institutions, and in assuming that these institutions impress upon each generation the sense of dependence upon the power of the State, of the commune, and of the family. And the more thorough-going historians attempt to explain how the protective

forms of these social institutions may have been determined by historical accidents of recent centuries. And
they make similar attempts to explain the spirit of independence of the British. Let us glance at three such
attempts.

Sir Henry Maine pointed to the great influence of Roman
Law upon French institutions; he showed how the French
lawyers, brought up in the school of Roman Law and holding
the Roman Empire as the ideal of political organization,
threw all their influence upon the side of the monarchy
and in favour of centralized administration. Some weight
must be allowed perhaps to this suggestion.

T. H. Buckle attributed the dominance of the spirit of
protection in France, partly to the influence of the Roman
Church, partly to the long prevalence of the feudal system
of social organization, under which every Frenchman was
made to feel his personal dependence upon the despotic
power of his feudal lord. This system, he said, culminated
in the despotism of Louis XIV with the subjection to the
king of the previously independent nobles. The dominance
of the spirit of independence in England he attributed in
a similar way to the character of English political institutions during recent centuries. After showing how, during
the sixteenth, seventeenth, and eighteenth centuries, the
people repeatedly succeeded in asserting its liberties
against the power of the kings, he wrote: 'In England the
course of affairs, which I have endeavoured to trace since
the sixteenth century, had diffused among the people a
knowledge of their own resources and a skill and independence in the use of them, imperfect indeed, but still far
superior to that possessed by any other of the great
European countries.' But he was not wholly satisfied with
this explanation, and added: 'Besides this, other circumstances . . . had, as early as the eleventh century, begun
to affect our national character and had assisted in imparting to it that sturdy boldness and, at the same time, those
habits of foresight and of cautious reserve to which the

English mind owes its leading peculiarities.' And that other circumstance which in Buckle's view was the primary cause of English independence was the establishment of the feudal system by William the Conqueror in a form rather different from that which it took in France. The English nobles received their lands directly from the king, and all landowners were compelled to acknowledge their obligation to the king. The nobles were in consequence unable to set up their own power against that of the king; and therefore, when they desired to resist encroachments of the royal authority, they called the people to their aid. Hence, he said, the English people early acquired rights and privileges and the habit of organized resistance to the central power. 'The English aristocracy being thus forced by their own weakness to rely on the people, it naturally followed that the people imbibed that tone of independence, and that lofty bearing, of which our civil and political institutions are the consequence rather than the cause. It is to this, and not to any fanciful peculiarity of race, that we owe the sturdy and enterprising spirit for which the inhabitants of this island have long been remarkable.[1]

A French historian, M. Boutmy,[2] recognizing and pondering the same difference of spirit between the French and the British peoples, would also explain it as the consequence of different political institutions obtaining since the Middle Ages. He points out that from the Norman Conquest the kings of England were invested with great power and were inclined to all the excesses natural to arbitrary power. Hence the first need of the people was to fortify themselves against the king. All the law of England, he says, carries the imprint of this fear and this defiance. The parliament has been set up against the crown, the judges against parliament, and the jury against the

---

[1] *History of Civilization in England*, vol. ii, p. 114.
[2] *Essai d'une Psychologie politique du peuple anglais*, Paris, 1903.

power of the judges; and so, ever since the Conquest, Englishmen have been accustomed to think, and to assert, that their persons, their purses, and their homes are inviolable; and that the State is an enemy whose encroachments must be sternly resisted, This way of thinking has by long usage become instinctive, increasing from generation to generation; until the horror of servitude has become rooted in the Englishman's nature and the desire of independence has become a native and primary passion of his soul.

These are interesting speculations, and their chief interest in the present connexion is that they illustrate the need of history for the aid of anthropology, the need of historians for anthropological studies. For these three distinguished historians write as though they accepted literally the biblical cosmogony and believed that the human race began with Adam and Eve six thousand years ago. They evidently regard one thousand or even five hundred years as a long period in the history of mankind. They do not seem to know that the races of men have been slowly evolving during hundreds of thousands of years, and that a period of a century or two is but a fleeting moment in the life-history of mankind.

I have shown you that both Buckle and Boutmy attribute the English spirit of independence to the fact that, less than one thousand years ago, England was conquered and ruled by a powerful despot; and that they attribute the opposite effect in France to a similar cause, namely the influence of despotic rulers. Surely their explanations are wholly inadequate! I will pass over the fact that these 'historical' explanations say nothing of the Lowland Scots, who have exhibited the spirit of independence in an even higher degree than the English, and will merely insist on the following considerations. If the English people had not already possessed in great measure the spirit of independence at the time when they were conquered by the Norman, surely his strong centralized rule would only have rendered them more dependent, would have fostered, in Buckle's phrase, the spirit of protection! If the characters of the French and British peoples had been reversed in this respect, how easy it would have been for these historians to show that the dependence of the English character was due to the crushing rule of the foreign despot, William of Normandy, and that the independence of the French was due to the existence among them in feudal times of many centres of independent power, the nobles, each capable of resisting the central authority! Surely, it was just because the spirit of independence was already theirs that the English people resisted their kings and were able to secure their liberties by setting up institutions congenial to their nature— institutions and customs which have fostered, in each individual and in each succeeding generation, the spirit of independence which they had inherited as a quality acquired by the race in the long ages of struggle and conflict before they emerged to the light of recorded history.

That these qualities, the spirit of protection of the French and the independence of the English, are innate racial qualities evolved during the prehistoric period is proved, not only by the impossibility of assigning any adequate causes operating during the historic period, but also by two other facts. First, by the fact that similar qualities are ascribed by the earliest historians to the principal ancestral stocks of the two peoples. It is proved also by the fact that other branches of the two stocks from which the Anglo-Saxons and the bulk of the French people are respectively derived have exhibited similar qualities, the same two qualities respectively.

Julius Caesar, Tacitus, and other early historians have described for us the leading qualities of the Gauls and of the Teutons. M. Fouillée[1] has brought together all the evidence of the early historians upon this matter. It

[1] In his *Psychologie du peuple français*.

shows clearly that the Gauls and the Teutons were distinguished by the same differences which obtain at the the present time between the French and English peoples, especially the difference in respect to the spirit of independence whose origin we are considering. I have no time to put this evidence before you, and I must merely mention the second great fact pointing in the same direction, the fact namely that other branches of the races from which the Anglo-Saxons on the one hand and the bulk of the French people on the other hand are derived have exhibited the same two qualities respectively, the spirit of independence and the spirit of protection. I refer more especially on the one hand to the Normans, the Dutch, and the Scandinavians, on the other hand to the Italians, the Russians, and the Irish.[1]

If, then, the historians have failed to understand and to account for this difference of nature, which has determined and is still determining so great a difference in the destinies of the French and British nations, are we to resign ourselves to ignorance? By no means? Let the historians seek the aid of anthropology and they may

[1] Further evidence of these racial differences is afforded by the spontaneous distribution in North America of European immigrants and their descendants. The English, Scots and Scotch-Irish (Scots by blood) have been the great pioneers of new territories. In the early days of European immigration a few Frenchmen of the type of Lasalle made wonderful journeys of exploration. But they were individuals commissioned by ᵗhe King of France to undertake such work, and there is reason to believe that these men were of the Nordic type. The mass of the French colonists have clung persistently to the eastern province of Quebec. The Irish and Italians, though largely of peasant origins, are to be found almost exclusively in the cities (discharging the highly sociable functions of policemen, bus conductors and local politicians) or in the closely settled agricultural regions. The Scandinavians, on the other hand, though late comers, have settled in the main upon the land in the wide open spaces of the west. The far west contrasts forcibly with the east and the more settled middle west (where German blood is much in evidence) in the marked predominance of the physical characters of the Nordic race. Jack London's *Valley of the Moon* is an interesting study at first hand of the profound effects in California of the racial peculiarities here lightly touched upon.

7

obtain new light on this and similar problems which have so intimate a bearing on their tasks.[1]

I should like to show you how one school of anthropology, the school of F. le Play, has attacked this problem, which the historians have falsely conceived and failed to illuminate, and how, in my opinion, they have gone far to solve it.[2] But time forbids. I may only point out that anthropology can not only assist in the solution of historical problems, but also demands a determining voice in many of the most urgent questions of statesmanship, where history alone, without the aid of anthropology, cannot afford the much-needed guidance.

History shows that the nations and empires and civilizations of the past have slowly climbed the slope of increasing culture and prosperity, have remained precariously poised some few years or centuries, and then have plunged steeply downward to stagnation or decay. Each one has described a parabola, or rather a trajectory, like that of a stone thrown from the hand. The supreme task of statesmanship is so to guide the nations of the present age that they may escape this fate, that they may

---

[1] It is not that historians and the many authors of studies of contemporary nations do not make use of supposed anthropological facts. Many of them (not less than fifty per cent I fancy) frequently offer explanations of their more interesting facts in terms of alleged racial peculiarities. The trouble is that they invent these peculiarities purely *ad hoc*, blissfully unconscious that they are laying down the law in a highly specialized field of science which they are utterly unprepared to deal with. In reading such authors I commonly amuse myself by making a list of the many weird 'instincts' which they invent for the explanation of the phenomena they describe. A few historians have explicitly recognized the need I am here insisting upon, but in theory only, not in practice. Thus Robert Flint, in his *History of the Philosophy of History* (1893) wrote: 'It is from the advance of comparative psychology that we may expect to see the most marked progress in the scientific interpretation of history in the near future.' Since these words were written forty years have gone by without bringing any advance in the direction so confidently indicated.

[2] A condensed statement of this solution may be found in my *Group Mind*, Cambridge University Press, 1920.

prolong the slowly rising curve, or at least may postpone
for as long as possible the onset of its decline.

History points clearly enough to the danger. But
history alone cannot tell us at what point of the curve
we stand; and it gives little or no indication of the causes
of the decline or of the measures by which it may be
postponed. Anthropology offers a theory of the decline
and suggests the measures that may prevent it.

I will try to state, in the few minutes that remain to
me, the view to which anthropology points more and more
strongly as the data accumulate and understanding grows.
I will embody the statement in seven concise propositions.
First, the races of mankind are of unlike natural endowments, some of them are better equipped than others
for the arduous task of creating and sustaining civilization;
and the same is true of the men and women of each race
and people. Within these walls, I state this truism with
some diffidence; for I know that this great college[1] is
distinguished by the breadth of its humane sympathies,
and I know that the truism is repugnant to our humanitarian sentiments. But it is our business to look facts in
the face without shrinking.

Secondly, only the better-endowed races and peoples
of the world are capable of developing or of sustaining
civilization of a high level.

Thirdly, those better endowed peoples are only capable
of producing and of sustaining civilization of a high level
in so far as they continue to produce in each generation
men of more than average natural endowments, intellectual
and moral, that is to say, men above the average level of
their own people, men of genius and talents of varied
kinds.

Fourthly, the peoples that can fulfil this essential condition are those formed by the blending of several races of
superior natural endowments and whose social institutions
are relatively free from 'caste'; in which, therefore, there

[1] Balliol.

exists in effective operation a 'social ladder' facilitating the rise of the talented and the descent of the incapable.

Fifthly, the operation of this 'social ladder' produces in the course of generations a stratum of the population which is richer in natural endowments than the rest of the population. In this stratum the native talents of the people become concentrated; and in each generation it produces a far larger proportion of men of outstanding abilities than is produced by the rest of the population.

Sixthly, for reasons which are easily intelligible and which are in the main psychological, this stratum tends to become relatively sterile. It ceases to maintain its relative numbers and is only kept up by constant recruiting from below.

Seventhly, this state of affairs, maintained for some generations, inevitably drains the whole population of its best elements; in consequence the people becomes relatively sterile of genius and talent.

When a people reaches that stage of its history, it is on the point of decline; and, unless there occurs some great change of social conditions which reverses the common tendency or somehow renews and freshens the strains of superior endowment, that people soon ceases to play a leading part in world history and rapidly declines to insignificance or extinction.

One of our most distinguished anthropologists, Professor Flinders Petrie, contemplating this process of the rise and fall of peoples, has denoted it by the phrase 'The revolutions of civilization'. I suggest that the nature of the process may be more accurately denoted by the words 'the parabola of peoples'.

The process implied by these words has been complicated in various ways in the history of each people—retarded in some, accelerated in others—by such factors as the infusion of new blood from outside, by emigration draining away superior elements, by conquests made or suffered, by the loss of superior elements by war or by colonization or by persecution of classes of superior endowments of special kinds. But the factor which has generally played the leading part and has probably been in all cases of considerable influence in determining the decline of peoples, the downward curve of the parabola, is what may be called the psychological infertility of the selected class. The formation of that class sustains the upward rise of the curve; its impoverishment brings the crisis; its further impoverishment makes the downward plunge inevitable.

In conclusion let me point to the practical bearing of the view which I have endeavoured to suggest to your minds in this one short lecture, without having time even to indicate the wealth of evidence upon which it is founded.

Many considerations point to the probability that our own nation is approaching, if it has not already reached or even passed beyond, the climax of its parabola. But let us not be paralysed by phrases and old saws such as 'History repeats itself' or 'The inevitable old age of nations', or by any other of the phrases in which dogmatic and unphilosophic determinism finds expression. Let us believe not only that man is free to choose within large limits his course of life and to secure success in it by the best efforts of his reason and his will, but also that nations are free in a still larger sense to choose and to pursue, and indefinitely to prolong, their course towards ever higher goals. Never before have nations had any such knowledge of the conditions of their success and of their failure as is now being placed at their disposal. This knowledge is a new factor in history which may be and, let us hope, will be of supreme importance in the future history of the world. For by its aid, and *by its aid alone*, we may hope to prolong indefinitely the rise of the curve of our parabola and make ourselves the masters of our national destiny.

If, then, any of you here are, or desire to become, historians, I beseech you that, while you are still young, you shall make yourselves competent at least to appreciate and evaluate the main results of anthropological research. Cease, I pray you, to regard anthropology as merely an inferior and crank variety of the despised 'stinks', a subject unworthy the attention of a self-respecting scholar of classical education. Call anthropology to your aid, and you will find your historical studies immensely enriched in interest and fruitfulness; then also you may aspire to rival your colleagues, the economists, in becoming the counsellors of statesmen and the saviours of your country.

VII
JAPAN OR AMERICA
AN OPEN LETTER TO H.I.M. THE EMPEROR OF JAPAN

SIRE,
    I am lately returned from a journey round the world, in the course of which I visited many lands, but none more beautiful and fascinatingly interesting than yours. I know not which to admire the more, the charm of the landscape, the noble productions of your great artists, the vast industrial progress made in so brief a period, or the friendly courtesy, the good taste, and the simple natural enjoyment of the beauties of art and nature so universally displayed by your people. During my brief sojourn in your land I learnt that you, the absolute ruler of a people so numerous, so ambitious, so energetic, so well endowed, are personally and actively interested in the problems of genetics, the science and art of improving the useful races of plants and animals. And at once I was filled with a new vision and a great hope for the human race. That vision and that hope I venture to express in the following pages, trusting that the sincerity of my motives and the importance of my topic will excuse in your eyes my audacity in addressing you.

    In all the lands I visited, I found much to admire and commend, but in all of them alike including both your country and my own, one reflection was repeatedly forced upon me; namely, we, the human race, are very ill-bred; ill-bred when compared either with the races of animals that live in a state of nature or with those which man has domesticated and modified for his own purposes. Among them one seldom sees a creature that is not graceful, healthy, efficient in all respects, full of vigour and vitality, beautiful according to its own type. How different is the lot of the human race! In every civilized land one sees among all classes a large proportion of men, women, and children burdened with defects of nature that derogate from their humanity, defects ranging from mere clumsiness of limbs or disharmony of features to gross deformities of structure. And among those who are physically passable one sees too many whose stolid faces and sluggish indifference betoken only too clearly that they have been denied man's highest prerogative, an alert intelligence, a capacity for any kind of higher intellectual activity.

Observing these lamentable facts, I reflected that the age which produced them is passing away, the age when peoples have contended by force of arms for 'places in the sun', the age when a nation was great in proportion to the number of armed men it could place in the field of battle. I reflected that we are entering upon a new age in which the rivalry between nations will no longer be conducted in terms of armies and navies, in which national success and glory will not be won on any battle-field; an age rather in which national ambition and national rivalry will succeed by achievement in the arts and sciences, in the art of right and beautiful living, of harmonious social organization, in the discovery of truth and its application to the enrichment and elevation of human life. In this coming age, great in this new rivalry, numbers will be of little advantage to a people. The quality rather than the quantity of a people will be the determining factor of success. No longer will any nation need great masses of 'cannon-fodder'. It will be esteemed great, not in virtue of great armies, vast ironclads and swift bombing planes, but by reason of absence of crime, disorder, injustice, fraud and cruelty, in proportion to the abundance of beauty, harmony, knowledge, wisdom and creative power, displayed in its collective and individual life.

Reviewing in my mind the peoples of the world, I asked myself—Which of the nations is destined to achieve the highest place in this coming age, the age of this new and higher rivalry? And there seemed only two possible answers—Japan or America. The nations of Europe are exhausted by war, hampered by jealously guarded frontiers and tariff barriers, burdened with old customs and deeply rooted prejudices, weakened by conflicts of classes and religions. They are populated beyond their natural means of subsistence and are so deeply engaged in the struggle to live that they have no energy to devote to the effort to live well. It is much if they can maintain such standards as they have, can avoid slipping downwards from the crest of a wave. America alone enjoys a superabundance of material prosperity.[1] Wealth pours from her soil and from beneath it. Her vast population bubbles over with energy and devotes itself to the production of more wealth with an enthusiasm almost religious. And no small part of the great surplus of resulting wealth is devoted to vast enterprises designed to raise and enlighten the mass of the people. America has left Europe far behind in the economic race; her people are enjoying all the immense advantages that wealth alone can give: and those advantages are very great, a high standard of public and private hygiene, abundant educational institutions magnificently equipped, museums into which are being swept the art treasures of the whole world, philanthropic agencies of every kind copiously endowed, ably manned, and organized to the last gaiter-button.

In many respects Japan resembles the European nations and is far out-distanced by America. The contrast between them is extreme. Japan is, judged by American standards, poor, poorer than many of the European nations. Her territory is insufficient for the support of her population without hardship. Her natural resources of fertile soil and minerals are small. Why then do I regard her as, alone among the nations, a serious rival of America in the new age of national emulation?

[1] Written in 1927.

In the first place and mainly, because the Japanese are a highly disciplined people. I do not mean merely that they have been drilled into obedience and orderly ways of living by a strong government. I mean rather that for long ages they have cultivated self-discipline, and that this self-discipline, directed by a conscious and noble patriotism, has in the modern age redoubled its efforts, and in so doing has rendered possible that wonderful transformation which all the world admires, the transformation within fifty years of an isolated, medieval, feudal peasant people into a great world-power, utilizing every discovery of modern science to enlarge and enrich its national life.

Secondly, the Japanese people, long accustomed to find delight and refreshment in the contemplation and production of things of beauty, has retained this priceless capacity in spite of all the modern changes; and, by reason of it, knows still how to live simply yet joyously, still knows how to combine plain living with high thinking and delicate feeling, still retains its fine taste in conduct and in art.

Thirdly, the Japanese people still practise assiduously and value highly the old pieties of hearth and home and temple and of personal loyalty to their ruler. For them religion and politics have suffered no unholy divorce. The old feudal devotions, which have adorned with countless instances of heroism the pages of Japanese history, have been concentrated in the one larger loyalty to their Emperor and their nation, inspiring men to self-sacrificing effort in the public services, to self-denial for the public good, in spite of all the allurements of the wealth-making careers opened by modern industrialism and finance.

Fourthly, Japanese culture is firmly rooted in traditions that have grown continuously through long ages, traditions which through their long process of growth and adaptation have become a homogeneous consistent system governing every detail of private and of public life, traditions sanctified by religion and by reverence for all that was great and noble in the past ages of the nation.

Fifthly, Japan, in becoming democratic, has retained that element of true aristocracy without which no democracy can long continue to flourish—the tradition of trusted leadership in all the great affairs of national life. The political life of the nation is guided by strong hands and brains, the hands and brains of men whom the people trust and follow because they believe them to be inspired by single-hearted devotion to the good of the whole people. This strong leadership and the people's trust in it have made possible the adaptation of the old traditions to modern life without destruction of them, without sapping their vitality and power to govern conduct in private and in public life.

In these five respects Japan enjoys immense superiority to America; and I hold that these superiorities suffice to offset the vast economic advantages of the American people, suffice to put Japan, alone among the nations, on a footing of equality with America at the outset of the new age of national emulation. For it is but too obvious to the most ardent admirer of America, to the most resolute optimist, to the most confident believer in the power of educational institutions to raise a whole people, it is too obvious, I say, that in all these five respects America differs profoundly from Japan. In the American people discipline is lacking, discipline of the home and the school, the discipline of religion and of the State, the discipline of self-conscious restraint and moderation in thought and feeling and action. Equally conspicuous by its rarity is fine taste in conduct and æsthetics; in spite of great achievements in architecture (which beauty is perhaps rather a by-product of the immense scale of buildings and of simple-hearted devotion to utility) the mass of the American people remain singularly indifferent to ugliness and to beauty and increasingly seek their satisfactions in those things which wealth can most easily furnish, vast concourses, lurid displays of crude magnificence, titillation of the senses, luxurious ostentation, speed, noise and perpetual agitation.

It is no less obvious that in America all the forces of tradition have been undermined and swept away, so that we have the spectacle of a great people groping in a moral chaos for some centre of gravity, some ideal, some governing principle, which shall take the place long filled in other peoples by religion and loyalty and devotion to the State; a people in whom political life has become trivial, the State a mere regulator of commercial prosperity, law merely an ineffective nuisance unrespected of the masses, true leadership unknown or scornfully and falsely envisaged as a form of slavery.

Turning now to contemplate the future, is it possible to foresee which of these two great nations is likely to attain pre-eminence? Which is of more value for the future —the economic advantages of America, or the cultural superiorities of Japan rooted as they are in immensely ancient traditions, unbroken and strong, yet adaptable through the influence of wise leadership?

The answer to this great question turns upon the answer to a narrower, more specific question, one which modern science can answer with no uncertain voice. Pre-eminence and the leadership of the world will accrue to that nation which first can learn to raise the quality of its population, to eliminate those crippling defects, those weaknesses and disharmonies of constitution, which, as I have said, are but too painfully obvious and widespread among all branches of civilized mankind. Who can doubt that any nation which in the course of a few generations can succeed in rendering all its members as strong, as vigorous, as intelligent, as public-spirited as its best individuals now are, that such a nation must have a glorious career, happy, wise and powerful, a beacon light for all the world, a career of world predominance in the new age of national emulation, going on to ever higher levels of achievement and self-improvement.

Now, if we compare Japan with America in respect of such possibilities of race improvement, we cannot fail to see that the realization of such possibilities may be far easier for Japan than for America. In America the tide sets strongly in the opposite direction. The fine old stocks that subdued the wilderness and laid the foundations of America's greatness are already well-nigh extinct. Their places are filled by a hodge-podge of stocks from many lands, stocks whose qualities remain to be proved by the demands of democratic citizenship. Among them the family, no longer rendered sacred by the sanctions of religion and patriotism, is losing ground more and more as the great school of character, the conserver of the social values and virtues, the foundation of the State. The high standards of comfort and luxury so assiduously sought and largely attained, especially by the women of America, are rendering the people soft, indisposed to undertake the heavy task of bearing and rearing children. The tradition of faithful discharge of primary duties has been destroyed; and women tend more and more to seek their satisfactions elsewhere than in the home and the tasks of motherhood. And for those to whom the rearing of a family still seems supremely worth while, everything conspires to discourage and hamper the execution of the task; the high cost of living, the difficulty in procuring domestic help, the multitude of amusements, engagements and social trifles, the pressure of professional ambitions, all the social customs conspire to make the young couples from whom sound progeny may be expected chary of giving such hostages to fortune.

The especial virtues of the American people conspire towards the same result. Their resolute optimism blinds them to their threatening danger of racial deterioration; their idealism forbids them to recognize the natural inequalities of men; their kindliness maintains a multitude of charitable practices and philanthropic institutions which favour the breeding of the strata at or near the bottom of the social scale, and prevents effective denial of the right to unlimited procreation to any couple however ill-fitted for the role.

Hence we see once more in America, the phenomenon, old as civilization itself, of a people dying away at the top, renewing itself from the bottom, increasingly leaving the tasks and the joys of parenthood to those strata of the population least capable of reproducing and maintaining the finer qualities that have made the nation great. And there is in America no power, no leadership, no body of instructed opinion, with sufficient prestige to enlighten and guide public opinion in this great question. Hence there is little prospect that the present trend may be arrested or reversed, that the ebbing tide of racial quality may be dammed back or made to flow once more to higher levels.

How much more favoured in this all-important respect is your great nation! A people among whom the cult of the family and of the ancestors has long been the keynote of its civilization, and among whom this powerful tradition has lost none of its vitalizing and preservative influence; a people long accustomed to the practice of severe self-discipline and self-sacrifice in the interests of the family, the community and the nation; a people whose religion makes little of the individual and much of the community; a people accustomed to look for leadership to the wisest and best informed; a people loyally and religiously devoted to the sacred person of its Emperor; and, above all, an Emperor who is both biologist and absolute ruler, who has at once the knowledge of the vital importance of racial qualities and the power to inspire and carry through measures for their conservation and improvement.

Is it not clear that you, the biologist Emperor, will call to your counsels all the wisest heads of your nation and will summon to their assistance from every part of the world the men who have made the profoundest studies of this great question? May it not well be hoped that such a council of statesmen and biologists, deliberating long and earnestly under your august inspiration, may evolve a scheme for the regulation of the reproductive process, a

scheme which the people, knowing it to be for the enhance-
ment of the welfare, the happiness, and the glory of the
nation throughout long ages to come, will accept with
enthusiasm and loyally observe, intelligently recognizing
its moral grandeur.

For the Japanese people has, I venture to think, recognized the fact that its future greatness must be qualitative
rather than quantitative; that it cannot equal its rivals in
numbers and sheer man-power, in magnitude of economic
production; that in these respects she must inevitably and
increasingly be out-distanced by the nations which possess
territories much larger than hers; that her future pre-
eminence must consist in breeding on a foundation of a
stabilized restricted population, well adjusted to its means
of support, the finest flowers of human culture, and in
providing them with the conditions most conducive to the
production of works of genius of every kind.

Unless, then, all the teachings of biology are at fault,
you have it in your power to initiate this new era of the
deliberate culture of human powers, to guide it on such
lines as may quickly bring the promise of magnificent
results. And you are of such an age that you may fairly
hope to see in your own lifetime the first-fruits of the most
enlightened effort yet made to raise a whole people to a
level of happiness and welfare and dignity vastly higher
and more secure than any hitherto attained.

May I, without presuming to forestall the labours of
your Grand Eugenic Council, suggest a scheme, which
though, in the light of deliberations of that council, it may
seem crude and relatively ineffective, may serve to aid the
imagination? Let all the children of your splendid system
of State schools be anthropologically surveyed, with the
application of every resource known to science; let the
family history of each child be minutely inquired into. In
the light of this survey let each child on leaving school or
college be given a grade mark indicating his biological
status as a member of one of six classes. Let each young

couple entering the married state be licensed to produce a prescribed number of children, a number corresponding to their position in the biological scale. Those of the lowest or sixth grade (few in numbers—for it would comprise only the markedly defective) would be sternly forbidden to procreate or, perhaps, rendered incapable of so doing by the harmless easy means provided by modern science—this as a condition of their marrying at all. Those of the fifth grade would be licensed to produce one child if they so wished; those of the fourth grade to produce two children; those of the third to produce three; those of the second to produce four (the two last-named grades would comprise the bulk of the population); those of the first grade would be licensed to produce at least five children and encouraged to produce more.

Americans would reject any such proposal as involving intolerable restriction on the liberty of the individual. But the Japanese people are wiser. They have long known that liberty to indulge every impulse without regard to social consequences is mere licence and is not essential to, does not contribute to, happiness; that rather happiness is to be found in such willing and enlightened self-discipline as contributes to the general welfare and secures to each man the respect and grateful recognition of his fellows.

VIII

THE ISLAND OF EUGENIA[1]

THE FANTASY OF A FOOLISH PHILOSOPHER

ONE of my children, after reading Mr. McKenna's story of the youth who inherited fifty million pounds sterling and who came to a miserable end, asked me, 'Dad, what would you do if you had fifty millions?' The question brought back to my mind a scheme which I had conceived in the enthusiasm of youth and had even committed to paper. That paper had lain forgotten for thirty years. Stimulated by my child's question, I found it among piles of unpublished manuscript. Having read it through, I decided that thirty years' growth of worldly wisdom had not made the scheme, conceived in all the ardour and ignorance of youth, seem any less desirable or less practicable. I have therefore recast my original draft, in the form of a dialogue between two men of middle age, whom I have named the Philanthropist and the Seer, or the Practical Man and the Scientist.

The Practical Man. My dear old friend, welcome to my summer retreat. I can hardly believe that thirty years have passed since we were chums in the old college, discussing so seriously our plans for reforming the world.

The Scientist. No, indeed! It seems but yesterday that we took our great decisions: you to devote yourself to the making of a great fortune, I to give myself wholly to the study of mankind; you with the conviction that science can do little without the power that money gives; I in

---

[1] This essay, which appeared in *Scribner's Magazine* (1923), was re-published as an introduction to the book, *National Welfare and National Decay*, and is reproduced here by the courtesy of Messrs. Charles Scribner's Sons.

the belief that men can easily be led to do the right thing, if only we have certain knowledge to guide them in the choice of what they shall do.

P. Yes, I remember it all. I did desire the power that money gives; and in all these years, during which I have been piling up the dollars, I have never forgotten my resolve to make good use of any wealth I might acquire. And now, as you know, I have made my pile, I have been successful. Good fortune, great opportunities, and good judgment have combined to make me one of the rich men of this rich country. I have arrived at the stage at which my wealth accumulates almost automatically, and for many months now I have been turning over in my mind the question—What shall I do with it? How can I best fulfil my youthful resolve?

S. Bravo, old fellow! I congratulate you. I had supposed you had gone over to the Philistines and sold your birthright for a mess of pottage. I can't tell you how delighted I am to find I was mistaken.

P. Thanks! But I want from you more than congratulations. I have persuaded you to come down here, because I want your advice. I want to take up our old discussions. I want you to advise me, to help me to make use of my wealth. You have spent thirty years in studying human nature and society. You have made a name for yourself in those studies. I want you to tell me whether, in the light of all your thinking and knowledge, you still have any of your old faith in the power of intellect and the good will to promote human welfare. Have you been disillusioned? Are you content to support your family respectably, to bear an honoured name in the academic world, to add a little to scientific knowledge, and then to pass away, with the vague hope that things will come right somehow, if only science progresses? Have not the terrible events of the last few years taught you that the increasing control of the physical resources of the world which science brings does but add to the difficulties and dangers of mankind? Isn't it clear now that civilization is in danger of destroying itself by the very means which science has so triumphantly provided? Is not the fate of Germany, its moral degradation, its political disorder, its economic chaos, is not all this a terrible warning? Does it not show that things of the sort you and I have been doing since our college days cannot ensure a better world, or save us from disaster? Germany excelled in our two lines of work, in the development of big business, and in the organized pursuit of science; and see what a mess she has made of her affairs. Her people were the most instructed, the best organized for peace and for war; her cities were excellently administered; her hospitals, her schools and universities were models for the whole world; her agriculture was scientific; only her churches were decaying. Does not the recent history of Germany show us only too clearly that all the things our philanthropists aim at, hygienic conditions, universal education, including popular interest in art and music, a rising standard of life, abolition of poverty, universal suffrage, that all these good things will not suffice to secure the moral health of a nation? Doesn't it look as though the mechanism of civilization which men have built up were getting too big and complicated for their control? When I contemplate putting my money into the promotion of any philanthropic scheme, the question rises in my mind—Will it do any good in the long run? Won't it merely accelerate a 'progress' already too rapid—I mean, of course, the increase, in scale and complexity, of the mechanism which is displacing culture?

S. I agree with you and I sympathize with you in your perplexity. The example of other rich men who have tried to harness their wealth to the service of mankind is not altogether encouraging. Carnegie's many millions are already beginning to seem a mere drop in the ocean. It is possible to doubt whether the world has been appreciably benefited by his gifts. Mr. Rockefeller has founded a great university and a great medical research institution. But the country can well afford to make and support its own universities; and it is thought by some of my friends that large endowed research corporations involve great drawbacks, even from the limited point of view of increase of knowledge.

Clearly, what you have to seek is some way of using your money which will fulfil two conditions: first, it must bring lasting benefit to mankind; secondly, it must be such as a rich democratic country will hardly adopt or support.

P. Yes, you define my problem exactly. My hope is that I may do more than put a plaster on some local sore, run soup-kitchens for starving millions, or, like Morgan, found maternity homes for deserving maidens. If there is no prospect that mankind, or some part of it, will achieve something more satisfactory than our present industrial civilization, then I would say—The more millions that starve, the better; and the less maternity, the sooner this miserable race of men will come to an end.

S. I see you are not an easygoing optimist. But I agree with you. You and your like should not devote your wealth to the applying of social plasters; nor should you give it for the benefit of institutions of the kind which should be and are supported by public funds. In the latter case your gifts would merely diminish in a hardly appreciable degree the rate of taxation throughout the country; and in a country so wealthy as ours the effect would be hardly worth considering. Nor should you devote your wealth to promote research either in the physical sciences, which are already enormously subsidized by industry; or in the medical sciences, for these can make effective appeal to every wealthy Philistine.

P. Well, you seem to have closed all possible roads for me. Is it really impossible to use great wealth to secure great and permanent goods? I have seen that proposition laid down most emphatically. But I have not been able to bring myself to accept it; and it is just because I don't see my way out of the difficulty that I am asking your advice. The people who make that statement are, I take it, the socialists or communists, those who think that all will be well if only private property can be abolished. And, upon my word, unless it is possible to find a way of spending wealth well, I don't see how its accumulation by individuals is to be justified. And yet, if we had a thorough-going communism, what would be the result? The masses of the people, especially the lowest strata of unskilled workers, would breed enormously, and this great country after a few generations of such breeding would be filled by hundreds of millions of low-grade population; we should become a second India. The game wouldn't be worth the candle.

S. I agree again. In that last remark you come near the essential problem. The only lasting benefit that can be conferred on mankind is the improvement of human qualities. Our social theorists propose all sorts of trans-formations of social and national organization, in the belief that mankind only needs to live under some particular ideal form of social organization in order to be for ever happy. The truth is that forms of organization are of little conse-quence; the all-important thing is the quality of the matter to be organized, the quality of the human beings that are the stuff of our nations and societies. Under the best possible organization of society, civilization will decay and go to pieces, if the quality of its human stuff is poor. Under the most anomalous social forms and faulty institu-tions, men will thrive and civilization will advance and improve itself, if the quality of its human stuff is sufficiently good. This is true on both the small and the large scale. The finest institutions will work miserably in incapable hands. Whereas, if your population is of sufficiently good quality, morally and intellectually, any institution will work tolerably; and in the extreme case, institutions and organizations, governments and churches may all be decently interred in favour of a happy anarchy.

P. Ah! I see where you are leading me. I begin to remember some wild scheme you talked about in our college days. An island to be called Eugenia, wasn't it, devoted to the production of supermen? It seemed to me, I remember, the wildest romantic nonsense. I didn't believe you were serious about it. Don't tell me that you are still hugging that fantastic notion in your middle-aged bosom.

S. Yes, indeed I am! Thirty years of study of man, of his history and institutions, have only confirmed my youthful conviction that such a scheme is profoundly worth while; that it is practicable; that the world is ripe for it, and needs it more urgently with every year that passes.

P. But look at the history of all such Utopian schemes. They all have fizzled out, or have been converted to ordinary humdrum industrial communities, after a very short time.

S. Yes, but they have all been run on wrong principles. You cannot argue that, because various imperfectly designed schemes of human betterment have failed, therefore every such scheme must fail. We are only now acquiring the knowledge that is essential for the wise designing of any such scheme. I will ask you to let me outline my plan and to give me the benefit of your criticism as we go along.

P. Very well, I'm ready to listen and play the critic. It's the least I can do in return for your willingness to advise me in my perplexity.

S. I begin, then, by stating the principles on which Eugenia is to be founded. Civilizations decay because they die off at the top; because, as they become increasingly complex, they cease to produce in sufficient numbers men and women of the moral and intellectual calibre needed for their support. So long as a nation produces in each generation a fair number of persons of first-rate calibre, it is safe; so long as its head is good, it can carry an enormous tail, without fatal decline. But the number of such persons tends to become not only relatively, but absolutely, fewer with each generation; because civilized societies breed from the bottom and die off at the top. I don't stop to substantiate this generalization. The evidence on which it is based is overwhelming. Instead, I will prescribe you a course of reading which will convince you of its truth.

The supply of persons of first-rate calibre can only be maintained by the fruitful mating of persons of superior strains. At present, as in all highly civilized societies of the past, such persons tend to be absolutely or relatively infertile. Eugenia is a scheme for bringing together persons of superior strains, for promoting fertile unions of a kind which will give to the world an ever-increasing number of persons of high calibre.

P. Then you wish to institute the human stud-farm *a la* Plato. Seeing that his scheme has been before the world more than two thousand years, why trouble to advocate it once more?

S. Not so fast. Plato's scheme involved the destruction of the family, the denial of conjugal affection and parental responsibility. No scheme which ignores the strongest tendencies of human nature can hope to succeed. Eugenia will avoid this fundamental error. Such neglect has been the ground of the failure of all the Utopias hitherto attempted. It will be founded on the cult of the family. Its religion will be something like ancestor-worship, tempered by a reverence for the progeny and by a great faith in their value to mankind. It is to be a place in which persons of superior strains shall come together in marriage and, under ideal conditions, produce the largest number of children compatible with the perfect health and strength of all concerned. It is to be an endogamous community, recruited by the admission of most carefully selected members and constantly improved by the extrusion of such of its native members as fail to come up to its standards of quality.

P. Then you propose to impoverish the rest of the world by bringing together in this community all its choicest spirits; no doubt in some such way an ideal community might be achieved. But it would be at the cost of the rest of the world. The essential selfishness of such a scheme condemns it to failure.

S. You are going too fast again. Eugenia is not to be ruled by a selfish regard for itself. It will be animated by the spirit of world-service. Its children will be brought up with the noble ambition to serve mankind. They will be true aristocrats; and their tradition will be *noblesse oblige*. The community will be a closed one only for the purposes of marriage and education. Membership in the community will be attained in every case by formal admission, after fullest inquiry into the family history and the intellectual and moral qualifications of each candidate. The advantages of membership, the attractions of life in Eugenia, the privileges of participation in its exalted aims, will no doubt attract many candidates from the outer world; and the best of these will be admitted. But, once it has become a 'going concern', Eugenia will recruit its citizens largely from the children born within its borders. Such children will not become citizens by right of birth alone. They also will attain membership only by formal admission. At the age of seventeen years they will become eligible; it may be supposed that the great majority of them will desire to become citizens, and that a large proportion of these will not fail to satisfy the strict requirements of family and personal qualification laid down by the fundamental laws of Eugenia.

P. Then, if Eugenia is to be a closed community only as regards marriage, it will not require to be a community dwelling within a territorial boundary. Its members may live where they please.

S. Yes and no. Eugenia must certainly have its own well-defined territory, a homeland over which it must exercise complete authority. To that question we will come back presently. At this point I want to define the relations of the members to the homeland. Since the people of Eugenia are to serve the world, they will be free to come and go, to dwell in other lands and to take up any honourable calling in those other lands. The only essential requirement is that they shall spend the years from five to twenty mainly, if not wholly, within the borders of Eugenia. After being educated in the family, in school, and in college, the young people will be encouraged to complete their education in the great universities of the world; only after doing so will they decide whether they will return to take up their life-work in the homeland or enter upon careers in some other country. And those who choose the latter course will not thereby sacrifice their membership. It will only be required of them that they marry within the community, that their homes shall be in Eugenia, and that their children shall be educated there. The relations of such members to the homeland may be illustrated by pointing to the relations of Indian civil servants to England. The Indian civil service has been a *corps d'élites* of Englishmen, who have accomplished one of the greatest tasks of recorded history, living and working far from their homeland. But they have not ceased to be Englishmen in the fullest sense of the word. They have married English women, their children have been brought up in England, their homes have been in England; and to those homes they have returned, when their years of service in India have been completed.

P. That's all very well. But the parallel fails in one important point. Your Englishman in India does not desire to marry an Indian woman. But your young Eugenians will go among people of their own race and of similar civilization. They will meet attractive persons, will fall in love and will marry and so be lost to Eugenia.

S. It is true that we shall probably lose in that way a certain number of our young people. But Eugenia can afford the loss, and that will be one of the principal ways in which she will be of service to the world at large. Her loss will be the world's gain. Every year she will contribute in this way to the populations of other countries a number of splendid specimens of humanity; youths and maidens perfect in body, of excellent moral disposition and character, and of outstanding intellectual capacity. But we may safely anticipate that we shall retain enough of the best to maintain the numbers and the quality of the community. The young people of Eugenia will be encouraged to look forward to early marriage, and they will not be prevented from doing so by economic or prudential anxieties. Every member will know that all his or her children, born from a lawful wedlock with another Eugenian, will be amply provided for and given the best opportunities for bodily and mental development that the world can provide. We may confidently expect that, before going abroad in the early twenties to complete their education, very many will be already engaged, and many even married. They will have been led to see that early marriage and the production of many children is their greatest privilege, at once their highest duty and their best guarantee of happiness. May we not hope that, under such favourable conditions, families ranging from five to ten children will be the rule rather than the exception, and that Eugenia will swarm with beautiful, strong, and perfect children, the delight and pride of their parents and the hope of the world?

P. But how about the women folk? We are told that the greatly restricted family, which has become the rule among the professional classes and the better-class artisans in all civilized countries is mainly due to the repugnance of educated women to become mere bearers of children and domestic drudges. Will not the highly educated young women of Eugenia take the same view and follow the same practice?

S. Undoubtedly some may do so, some in whom the maternal instinct may be weak or who for any other reason may fail to absorb the ideals of Eugenia. And these will go out into the world and will not return. But the moral atmosphere in which the girls will grow up, and the high esteem in which parenthood will be held, the appeal to all that is best in them, will prepare the majority of them to face with enthusiasm the sufferings, the trials, the sorrows and the joys of motherhood. And as for domestic drudgery, the whole plan of life in Eugenia will be directed to diminish to the utmost the more mechanical and menial of the domestic tasks. The mother of a large family will be aided in a hundred ways; not only by perfection of household arrangements, but by having helpers who will find their highest happiness in such work. Grandmothers, widows, spinsters, all the women who have no young children of their own to care for, will give at least a part of their time to helping the mothers. And so the young mother, instead of being worn down to premature old age by anxiety and drudgery, will find abundant helpers, educated gentlewomen like herself, to whom she can entrust the partial care of her brood with perfect confidence.

P. I begin to think there may be some sense in this fantastic scheme of yours. Assuming that you have secured a fine stock of human beings to begin with, you count on their multiplying at a natural rate under the favourable moral and material conditions you hope to provide; from this natural increase you discard constantly the least fit by denying them membership in your community; and you seek further to improve the stock by admitting from time to time a certain number of highly selected persons.

S. Yes, that's the essence of the scheme. It is to secure all the advantages of a most rigid selection, not by the cruel methods of nature, whose great instrument is the selective death-rate; but by a purely beneficent selection, which shall substitute for the death-penalty merely deprivation of the right to marry within the community, or, more strictly, deprivation of the right to remain within the community to all those who undertake marriage or procreation in defiance of its laws. And, by admitting new members selected from the whole world according to the strictest principles, the immense benefits of rigid selection within the stock may be indefinitely augmented. Every biologist will tell you that, if such a scheme can be worked for a few generations, you can count upon producing a remarkably fine stock. Whether we can hope to secure in this way the production of many men and women of the first order of intellect, actual giants or geniuses, that is a further question, the answer to which is open to debate. But there is no room for doubt that we may expect to have a stock almost every child in which will be fitted to attain eminence in some walk of life and to render great services to his fellow-men.

P. Well, I'll grant, for the purpose of this discussion, that you may fairly anticipate this magnificent result if, as you say, the scheme can be worked. But what a large 'IF'! How are you going to start it? How bring together your choice spirits, the new Adams and the new Eves? And, if once started, how can you hope to guarantee it against the fate that has swiftly overtaken every other community of cranks, namely, dissension and dissolution.

S. I grant you that most of the many crank communities have had short lives and have been fit objects for the world's derision. But let me point out that there is at least one exception, namely, the Mormon community. All those others, from Brook Farm to the Oneida Settlement, have suffered the fate they deserved, the fate that might have been confidently foretold by any one with a little knowledge of human nature; for they ignored or defied the fundamentals of human nature. The Mormons, on the other hand, have flourished greatly and have achieved a community which in very many respects outshines all competitors. That is because their fundamental principle was in accordance with an outstanding, an undeniable, fact of human nature, the polygamous tendency of the human male. Now, as I have said, Eugenia is to be founded on monogamy. For, though man is polygamous, woman is not. Woman, when her nature is unperverted, prefers to have her own man and her own home and her own children about her. And, unless it could be shown that the biological welfare of the group absolutely demands polygamy, the ideal state must demand of man that suppression of his polygamous tendency which the happiness of women, the stability of the family, and his own spiritual welfare alike require.

P. I had forgotten the Mormons. I admit they are an exception to the rule. But let me hear how you propose to initiate your great experiment.

S. This is where you come in with your millions. Other rich men have given vast sums to endow universities, libraries, research-institutions, peace-prizes, and so forth; none of these, as you yourself have so clearly seen, promises any lasting benefit. I offer you now a means of applying your wealth in a way which, in its promise, far surpasses all these. If you adopt my plan, you may feel a reasonable confidence in the attainment of great results. And, when at some future time you and Andrew Carnegie look down upon this world from some distant star, you will point out to him with legitimate pride the sons and daughters of Eugenia, and will unroll before his envious eyes the record of their great achievements.

> Ah! make the most of what we yet may spend
> Before we too into the grave descend.

The first essential is a suitable territory. An island would have many advantages. But, whether an island or not, it should be not less than five hundred square miles in extent, and might with advantage be as large as ten thousand square miles. It should have a white man's climate and a reasonably fertile soil. It should have some natural beauty. The more diversified and beautiful its natural features, the better it will serve our purpose. There you are to found a great university for both sexes, the nucleus of a group of great professional schools and centres of post-graduate study and research.

By offering adequate salaries and ideal conditions of

living and working, you will attract a brilliant staff of instructors. In appointing them regard will be paid, not only to the personal qualifications of applicants, but also to their family histories; and, in the cases of those who are already married, the personal and family qualifications of wife or husband will be taken into account. In other words, your instructors will be persons not obviously disqualified for membership in the community, but rather *prima facie* qualified for admission. After some years of service, the question of their admission to full membership will arise, and each case will be carefully considered on its merits. He (or she) who does not desire full membership, or has not the necessary qualifications, will resign his post and transfer his talents to some other sphere. Those who desire full membership and who have the required qualities will be solemnly admitted. Whether they will suffer rites and mysteries at their initiation, the taste of the community may decide; but some formal and public recognition of their admission must be made. In this way you will secure a nucleus of Eugenians. This nucleus will grow rapidly by natural generation; for marriage and parenthood will be held in high honour; and everything will be done to make smooth the path of family life. The brilliant young teacher or writer or scientist will not be torn by the conflict between his ambition, his desire to do great work, his sense of the supreme value of the intellectual life, and his natural desire for marriage, a home and a family. In this happy land duty and inclination will coincide; the two chief goods of human life will not be incompatible. Failure to marry and sterility after marriage will be serious bars to the continuance of tenure of position; the bachelor and the childless will enjoy no economic advantages over those whose quivers are full.

And this nucleus will be recruited by many of the best among the students who will flock to your university from all parts of the world, attracted by its unique advantages. These also will be selected from the swarm of applicants with due regard to racial and family, as well as personal, qualifications. And, when they shall have passed one, two, or more years within the walls of your university, they will become eligible for membership in the community. They will probably have to serve a probationary period after election.

P. Then you propose to create a community of lily-fingered scholars, living on imported tinned meats, jam and pickles, a swarm of parasites, not one of whom could do a decent day's work or earn an honest living, if thrown on his own resources.

S. My dear sir, when did you acquire this exaggerated respect for the horny-handed sons of toil? Eugenia will be in a certain sense and degree parasitic; just as every university, every institution of the higher learning, of art or of science, is parasitic, as you and I are parasitic; in so far, namely, as the Eugenians will not be wholly or chiefly employed in securing the primary necessities of life. But, like those other institutions, and in a far higher degree, Eugenia will justify its parasitism by the great services it will render to the world. Symbiosis, rather than parasitism, is the right word to use in describing its relations to the world at large. But granted a partial dependence upon the outer world as inevitable, Eugenia will not forget how Adam delved and Eve span. Every child will be taught a trade, will learn how to use his hands as well as his head; and, when he grows to man's estate, he will not need to relieve the tedium of his leisure with interminable rounds of golf or sets of tennis. Games will not be taboo; but every man will be expected to devote a part of his time to practical labour of an immediately useful kind. The professor of astronomy will milk the cows; the expert in chemistry will handle a Fordson tractor with a satisfying efficiency; the social anthropologist will look after the bee-hives; while the professor of fine arts or philosophy may be expected to cultivate his sense of humour by attending to the pigs. I think we may hope that, when once the little State shall have got into smooth running order, it will become nearly self-supporting, will rely more and more on its own economic activities and marketable products and less upon its endowments.

P. And how about the political status of Eugenia? Do you see it under its own flag, maintaining its own army and navy, and sending out its ambassadors to lie abroad?

S. No, certainly not. It is clear, I think, that it will need the protection of a great power, and that it must fly either the Stars and Stripes or the Union Jack. It might well be a territory of U.S.A., or a British Protectorate. Sarawak might serve as a model in this respect; a little country, about as big physically as England, ruled most happily by an Englishman and a small handful of colleagues, but protected from aggression by all the might of the British Empire, while free from all interference by the British Government.

P. Then Eugenia is to be as nearly as possible an independent State, having its own laws and political constitution.

S. Yes, that is the most desirable status for it. The constitution might be very simple and the laws few. Eugenia should approximate to that happy condition dreamt of by the philosophic anarchists, in which courts of justice and laws and police for their enforcement are all outgrown. But I recognize that it may be difficult to secure for it a suitable territory not already subject to the laws of one of the great Powers and you might, I think, be content to see it subject to the laws of either U.S.A. or Great Britain. I can see no harm in that, so long as the community owns and controls its territory. In the beginning its territory and public buildings and financial resources should be controlled by a board of trustees; which board might well be recruited as time goes on in accordance with the principles of representative democracy.

P. If your plan is successful and your Eugenians multiply freely, as you wish them to do, they will double their numbers every twenty years or so; and so your little State will soon be threatened by the spectre that dogs every successful State, the spectre of over-population; quite apart from those desirable new recruits whom you design to attract from all parts of the world.

S. There again Eugenia will be in a peculiarly happy position. The more her people multiply, the more she will feel she is playing her part in the world well and truly. As soon as the community shall have attained to such a number as seems the optimum for her territory, she will need to retain in her service only two members of each family of children; the others, three, four, five, or more, will be launched upon the world to seek their fortunes. They will go out, splendidly equipped in mind and body, with noble ideals of service, to play their part in the great world. No doubt many a one will tear himself away with a heavy heart; just as many a youth has found it difficult to leave the happy scenes of his college life. But, the step once taken, they will soon make for themselves honourable careers. They will be the salt of the earth, leaders in all the professions, bound together and to their Alma Mater by memories of their happy youth and by their sense of their part in the realization of a great ideal. And this surplus increase of population will give to Eugenia the opportunity of a sustained and stringent selection which, if wisely used, will result in a continued evolution of human qualities to which our reason can foresee no limits and which our imagination cannot depict.

P. I foresee that your plan will require all of the fifty millions sterling of which you spoke at the outset of our talk, if it is to be adequately launched and endowed. I must put it before some of my fellow dyspeptics who, like myself, can't stomach all their wealth. Meantime, do you work out your scheme in greater detail and subject it to the criticism of your colleagues. Get together a group of biologists, psychologists, sociologists, and people of that kind, and let them pull it to pieces if they can. Then we will meet again

9

and have another talk on ways and means. But one last objection before we turn in! Obviously your scheme will encounter the derision of all Philistines. I don't know that that need trouble you. But is there in the world any considerable number of persons of the right sort, persons who would uproot themselves from the well-trodden pathways in order to venture their lives in the crazy bark imagined by you? I mean is there any hope that a group of reasonably well-endowed persons could be got together to initiate your colony? Won't you have to be content with a collection of cranks and failures, ill-balanced visionaries and discontented paranoids, seeking refuge from a world that has proved too hard for them?

S. I haven't the least misgiving on that score. There are thousands of young people asking themselves at this moment: What is best worth doing? How can I devote myself to some course of life that is really worth while? You and I, my dear fellow, when we burned with the ardent desire to do some good in the world, were not rare exceptions. If only you and your fellow dyspeptics will do your part, Eugenia can be made so attractive, so appealing to all that is best, all that is idealistic in human nature, that we shall practically be able to select our members from the whole human race. Candidates for admission will surge around our doors. The deliberate founding of a colony is not a new and unheard-of-proposition. Think of Greece, of England, of New England, of Virginia, of New Zealand. Perpetual colonization has been the essence of the history of the Nordic race. That fine race seems in fact incapable of surviving, when it ceases to migrate and colonize. It is one of the virtues of Eugenia that it offers the prospect of saving a remnant of that disappearing race, of perpetuating the stock and restoring and perhaps even enhancing its ancient virtue. But I don't want to raise the racial question, with all its inevitable prejudices. The effects of race-blending and the many allied problems of human biology will be a principal field of research in the

University of Eugenia. We may confidently expect that this department will attract the most brilliant of our students. Historical study will no longer mean minute research into ancient charters and forgotten personalities; it will be a sub-department of anthropology. The science of man will for the first time receive adequate recognition; that is to say, it will dominate the scene. To it all other sciences will be duly subordinated; and they will be valued and studied in proportion as they contribute to the solution of its supremely important problems. For the foundation of all the activities of Eugenia will be the conviction that the human race, if it is to have a future which we may contemplate without horror and despair, can achieve such a future only by deliberately assuming the control of its own destiny.

IX
FAMILY ALLOWANCES[1]

IT may be assumed that the term Eugenics, as introduced and used by Mr. Francis Galton, is familiar to all
members of this Society. I may also assume that we should
all agree in wishing to promote any measure, any change
of custom or institution, which could be clearly shewn to
be eugenic in tendency and to involve no injustice, no interference with personal liberty, and no risks of weakening
or destroying any of the pillars of our social system. The
suggestion I wish to put before you is the desirability of
a change of custom which would fulfil those conditions.
And the bearing of my remarks may be clearer if I at once
define this proposed change of custom.

There are in this country certain large classes of persons
selected from among the whole population by tests which
ensure that in the main these persons have a civic worth
above the average. My suggestion is that we should
endeavour to introduce the custom of remunerating the
services of every person belonging to any such selected
class, not, as at present, according to some rigid scale, but
according to a sliding scale such that his income shall be
larger in proportion to the number of his dependent offspring.

The agencies by which these persons are selected are
constantly becoming more efficient and more wide-reaching, but they probably operate in the main as agencies
of degradation of the population, through making against

[1] Read before a meeting of the Sociological Society, at the School
of Economics and Political Science (University of London), Clare
Market, W.C., on February 21st, 1906, and published in *Sociological
Papers*, vol. ii, under the title 'A Practicable Eugenic Suggestion'.

the rate of reproduction of the individuals selected by them. A change of custom of the kind suggested would convert them to effective eugenic agencies favouring very greatly the reproduction of the selected classes, classes which comprise a large proportion of all the individuals of more than average civic worth. But the main contention of this paper is that the suggested change of custom may be expected to favour very greatly the reproduction of the individuals and classes of *highest civic worth*, and that, if it should have this effect, it would be a eugenic influence of vastly greater importance than the negative measures so commonly advocated.

This contention is based upon the three following propositions : (1) that some men are of very much greater civic worth than the average, and that the continued strength, prosperity, and progress of this or any nation depends upon the continuance of a good supply of these persons of high civic worth ; (2) that mental and moral qualities are hereditary in much the same sense and degree as physical characters; that, therefore, the superior elements of the population in each generation, and especially the persons of highest civic worth on whom the continued welfare of the nation is mainly dependent, will be found in far larger proportions among the progeny of the superior individuals of the preceding generations than among the progeny of the mass of persons of average qualities; (3) that in this country at the present time, the fertility of the superior classes, or, better, of the individuals of higher civic worth, is low relatively to that of the mediocre mass of the population, and still lower relatively to their maximal natural fertility, and that this relative infertility is mainly due to artificial and removable causes.

Following Mr. Galton, we may call *illustrious* the great men who give to the world ideas of supreme importance. Now, while it is true that the progress of mankind in general is in very large part due to the activities of these illustrious men, it must be admitted that for a space of years, probably for some generations even, a well-organized nation might continue to be vigorous and healthy and to hold a good place in the world, though it should fail to bring to maturity any man of illustrious powers. But this would be possible only if it continued to produce in considerable numbers personalities of what we may call the second order of capacity; men who, though they are not endowed like those others, like Newton or Wordsworth or Bentham or Darwin, with powers that enable them to set going new movements in the world of thought or action, are yet full of intellectual and moral energy of a high order; men of the order of ability that we may roughly define by imagining grouped together some fifty of the most capable and efficient members of the houses of parliament, and the corresponding fifty from each of the great public services and from each of the great professions and callings. Such a group would correspond to those whom Mr. Galton classes as eminent men; and of such eminent men, and of younger men whose capacities are such as to raise them to eminence in middle age, he reckons that this country can boast about two thousand at the present time. The sum of the services rendered to their country by these eminent men cannot be reckoned inferior to the services of the illustrious men, and indeed they should perhaps be reckoned of more importance; for the ideas created by the men of supreme powers are given to the whole world, or to all that part of the world that is capable of appreciating them, and so advantage but little in the international struggle the country that gives them birth. But in the absence in any country of a fair supply of the minds of the second order, the great gifts of the world's illustrious men must remain ineffective in that country; for it is they who mediate between these moving spirits and the great mass of mediocre men, interpreting and teaching to the latter the ideas originated by the former; and it is they who maintain by their thoughts and conduct the highest traditions of the national life. If we try to imagine all, or a considerable proportion, of the men of this second order simultaneously removed by death, we may realize something of their importance; for it is clear that in the course of a few years from that event the nation would be reduced to a state of moral, intellectual, æsthetic and social chaos. It may be laid down in general terms that, while on the one hand, the world's illustrious men are the source and cause of the progress of mankind in general, in all that is worthy of the name; on the other hand the continued prosperity, stability and vigour of any nation is chiefly dependent upon the production in sufficient numbers of men of the second order of capacity.

Our greatest authority on all such questions is Mr. Galton, who for many years has devoted his great talents to the study of them. His conclusions received their most definite expression in the Huxley Memorial Lecture of 1901, entitled 'The Possible Improvement of the Human Breed under the Existing Conditions of Law and Sentiment'. Mr. Galton imagines the whole population of the United Kingdom divided according to their degrees of civic worth into a series of ten classes, the classes being so defined that the difference between the average civic worth of any two adjoining classes is equal to that between any other two, *i.e.* the ten classes form a scale of civic worth rising by steps of equal value from the lowest to the highest. The two classes occupying the mid position in this scale together comprise one-half the population, and in numbers, and probably in civic worth, correspond fairly well to Mr. Charles Booth's class of artisans earning from twenty-two to thirty shillings a week. The five classes of individuals of more than average civic worth, Mr. Galton denotes by the letters R, S, T, U, and V. These five comprise about half the whole population, and of them the lowest, R, the class of persons just above the average, comprises about one-quarter of the total population; S comprises about one-sixth, while T, U, and V, the three highest classes, together comprise about one-tenth only of the whole.

Mr. Galton shows reason to believe that, if all these classes were equally prolific, the V, or highest, class, would be three times as rich in V-class offspring as the U class, $11\frac{1}{2}$ times as rich as the T class, 55 times as rich as the S class, and 143 times as rich as the R class; so that in spite of the smallness of its numbers, class V would be *absolutely* many times as rich in V-class offspring as class R, together with all the five classes below the average; and further, that the classes T, U, and V of any one generation, comprising together only about one-tenth of the whole population, might be expected to produce four-fifths of all the V-class individuals of the succeeding generation.

Here it is necessary to point out that the argument does not assume that the classes of Mr. Galton's scale correspond strictly to any of the commonly recognized social grades. Our social grades are of course based largely on wealth or income; and as wealth is hereditary, it follows that many persons continue to enjoy a far higher social standing than is warranted by their civic worth and that the higher social grades are therefore very mixed, i.e. contain large numbers of persons who in the scale of civic worth belong to the mediocre or to some lower class.

Nevertheless, it remains true that the higher social classes, especially perhaps the class which we roughly define as the upper middle-class, and which comprises most of the intellectual workers of the country—the members of the higher professions—is the product of a long-continued process of selection. For many generations, the ablest members of the working classes and lower middle-class have been able to emerge from their class and to establish themselves and their families in a higher social class; and conversely, though perhaps to a less extent, the least capable members of the higher social classes have been falling back to lower social grades. And it may be claimed, I think, that we have now well-nigh perfected the social ladder. Free and compulsory education; abundant, perhaps too abundant, scholarships; the breaking down of all rigid social barriers; the general tendency to seek and promote capacity wherever it may be found, and even a certain sentimental tendency to exaggerate the merits of the man who shows a tendency to rise: all these together ensure that now, and in the future, ability, or rather civic worth, can find its opportunity and its appropriate social level; and they justify us in believing that it seldom remains hidden. This is true in spite of the fact that birth and wealth may, and often do, favour the success of able men born in the upper classes.

We may, then, believe that now and in the future the individuals whose innate capacities render them members of classes T, U, and V will be found occupying approximately such positions in society as their superior merits entitle them to claim. These persons constitute, as we have seen, about one-tenth of the population, and we may fairly regard them as an emerged tenth, a tenth emerged in the past or emerging now from the mediocre mass, through merit.

These upper classes, T, U, and V, which, if equally fertile with the remaining classes, would produce many more persons of eminent civic worth than all these remaining classes taken together, are at the present time relatively infertile; and the principal causes of this infertility would seem to be artificial and removable.

These principal causes are two: namely, late marriage and voluntary restriction of the size of the family after marriage. The custom of late marriage tends to diminish in two ways the rate of reproduction of any class among whom it obtains. If in two equally numerous classes an equal number of children is produced on the average by each pair of parents, but the one class marries on the average at the age of twenty-five years, while the other class postpones marriage to the age of thirty-five years, then the generations of the former class will succeed one another so much more rapidly than those of the latter class, that after a few generations the numbers of the former will far exceed those of the latter in spite of their equal fertility. But, secondly, the offspring of late marriages will, other things being the same, be less numerous on the average than the offspring of early marriages. That voluntary restriction of the number of offspring is peculiarly and increasingly common among the well-educated and intellectual classes will, I think, be generally admitted; although it is of course impossible to produce statistics bearing on this question and to separate the effects of voluntary from a possible natural infertility.[1]

These two influences, late marriage and restriction of the family, are at a maximum among just those members of the upper social strata who constitute our emerged tenth. For these persons are chiefly to be found leading the strenuous life in the higher intellectual professions, and especially in those callings to which access is gained only by success in intellectual competition. And the more ambitious a man is, the more he is engrossed in his work; and the more highly cultivated and naturally keen are his tastes and intellect, the more likely is he to remain a bachelor, or to marry late and to restrict the number of his children when married. On the other hand, many of the social changes which have been recently effected or are now going on in this country directly favour the reproduction of the inferior classes. Such are the low price of bread and sugar, the tendency to throw taxation chiefly on the well-to-do classes, enormous charities, free medical and surgical treatment, free education, free feeding of school-children, free milk-depots, the building of dwellings for the working classes out of public funds, and lastly the abolition of the excessive infant mortality among the lowest classes.

At this point it seems desirable to emphasize the fact

[1] The fact that there is some ground for suspecting that the intellectual classes are naturally less fertile than the others, does but add to the urgency of the need for the reform of custom suggested in this paper.

that the effects of different rates of reproduction of different classes must be cumulative in a surprising degree, if the difference is maintained through several generations. Professor Pearson has dwelt upon this point in his essays upon 'The Chances of Death'. He points out that if in any community one group of people having a fertility above the average breeds in from generation to generation, i.e. if its members intermarry only or chiefly with members of the same group, then this group will tend rapidly to replace the other groups of the population; so that, after a surprisingly brief period of time, almost the whole population will be descended from the most fertile group. And he shows that changes of this sort may well be going on in this country, because fifty per cent of the people of each generation are produced by about one-fifth only of the adults of the preceding generation. If then the classes of highest civic worth reproduced themselves equally or more rapidly than the mediocre classes, there would soon be established a breed capable of producing in each generation a very large number of persons of eminent abilities, because not only would these higher classes be continually recruited by the most able offspring of the mediocre classes, but the least able and worthy of the offspring of the highest classes would constantly fall back to become members of the mediocre classes. On the other hand, when—as is almost certainly the case at the present time in this country—the classes of highest worth are persistently less fertile than the mediocre classes, then they must be recruited on a far greater scale by persons drawn from those mediocre classes, on such a scale that the average worth of the highest classes may be dragged down to a lower level. There is, then, always the probability that in the course of a few generations any strains of exceptional ability, such as have undoubtedly appeared from time to time, will, through intermarriage with much inferior strains, regress markedly towards the mediocre type and become in fact swamped or extinguished, ceasing to produce any notable proportion of eminent men, and serving merely to raise in an infinitesimal degree the general average of ability throughout the nation.

I have now indicated the principal lines of reasoning which justify the statement with which I set out; namely, that, at the present time in this country, it is of far greater national importance to promote, if possible, a higher rate of reproduction of certain superior classes than to provide for the elimination of the unfit. The considerations advanced justify us in believing that, to secure by the latter method benefits to the national breed at all comparable to the benefits which may be confidently expected from a successful effort of the former kind, we should have to immolate or isolate vast numbers of the least desirable specimens—numbers so large that the population of the country would be seriously diminished—so that from the point of view of the national power and prosperity in the present, the cure might well prove worse than the disease.

The facts are sufficiently well established to convince all intelligent men of this nation's urgent need of such changes of custom, law or institution as will tend to promote the reproduction of the superior elements of the population, and in fact of all classes which may reasonably be regarded as above the average of civic worth.

It seems worth while at this point to consider the limits within which such changes must be confined. In the first place, I submit that they must not be such as to undermine or destroy the institution of the family; because, as I in common with many others believe, every great and stable civilization has been based upon, and can only be based upon, a sound family-life. Hence the methods of the stud-farm, though advocated by Plato or by Mr. Bernard Shaw, or by any other equally distinguished writer, must be ruled out. This, however, leaves open the question of the form of the family; and much can be said in favour of a restricted polygamy (not the harem). For I do not think that any proposed change is bound to be consistent with the existing state of law and sentiment. Law and sentiment can easily be changed if sufficiently good reasons can be shown; and it is from such changes that we have most to hope. Something may be hoped for from a wider diffusion of a knowledge of the conditions, especially among women. For it may be supposed that, when the conditions are made clear to all, the really superior women will cease to regard as their principal duty either the attendance at, or the engineering of, social functions, philanthropic or otherwise; that they will once more find their highest duty and pleasure in producing, rearing, and educating the largest number of children that their health and their means will allow; that they will realize that this is a career and a profession, difficult, interesting and honourable in the highest degree, compared with which the careers and professions followed by the great majority of men, even the most successful men, are dull, stale and unprofitable both for themselves and for society.

But an important and, as I believe, by far the most important cause of the relative infertility of the better classes, and especially of those among them whom I have called the emerged tenth, is the consideration of income and expenditure. For the current market-price of good highly trained abilities, i.e. of the services of educated men of abilities of our classes T, U, and V, lies between £500 and £1,000 a year, say about £700 a year. This is just such an income as tempts a man of highly educated tastes to remain a bachelor, to postpone marriage, or to restrict severely the number of his children when married.

If this is true, then we may hope something from the spread of a tendency of the best people to mark themselves off from the common herd by a resolute rejection of the wasteful and uselessly luxurious habits of life that have become so common among us in recent years; by the rejection of the 'champagne-standard' in fact, and by the practice of a simple mode of life which, while not despising luxuries of the better sort, knows how to discriminate between them and mere ostentation.

But I think we may hope for still greater results from the general adoption of the change of custom that was briefly suggested at the outset of this paper. It is convenient to illustrate the influence of income and to show how the reform I suggest may be introduced by considering the case of a single highly selected and salaried class; namely, the civil servants.

At the present time the State not only does nothing to promote a relatively rapid multiplication of the intrinsically superior elements of the population, but it actually maintains an extensive and unjust system by which it restricts the multiplication of those elements. The State has in its pay a large number of public servants, all of whom are selected from among many competitors, and of whom a considerable number, namely the first-class clerks of the home civil service, are selected from among the ablest youths of the country, by very severe tests of mental capacity and physical soundness. No one will deny that these constitute a group of men of high average capacity. They are selected by carefully conducted competitive tests from the ablest youths of our universities, chiefly Oxford and Cambridge, to which in turn most of the ablest boys of all the schools in the country find their way. In order to secure a place in this service, a youth must have not only great mental capacities, but also moral qualities of no mean order, namely, an energy and steadfastness of purpose which enable him to apply himself steadily and effectively to the education of his powers and to the acquisition of learning throughout the years of school and college life. And the competition is so severe and the age limit for candidates is so rigidly drawn that success in the competition implies, save in cases of very exceptional ability, the possession of a sound constitution; for any youth who through ill-health is prevented from continuous and steady work at school and college inevitably finds himself unprepared for the competition when he arrives at the prescribed limit of age. The civil servants of the first-class constitute, then, a very highly selected group, of which the average civic worth is probably not less than that of the V class in the scale of ten classes described above. They constitute, in fact, a class of which the offspring may be confidently expected to contain a relatively large proportion of persons of eminent capacities. It is then of high importance to the State and for the future progress and welfare of the nation that these men shall produce a reasonably large number of children. Now these men receive on entering the service a salary of about £200 a year, and the salary increases on continued service up to a sum which varies from about £600 to £1,000 a year, except in the few cases of those who, becoming heads of departments, receive as much as £1,500 or £2,000 a year. We shall not be far wrong if we say that on the average the salary rises from £200 to £800 a year and put the average at £700 a year. This is an income which the ordinarily successful business man would regard as pitfully small and miserably inadequate for the bringing up of a family; nevertheless, the career offers compensating advantages, and since the salary, in conjunction with these advantages, suffices to attract as candidates for the service large numbers of the ablest young men at the universities, it must be regarded as adequate, or, at least, the State cannot well be called upon to make any increase in the amount of this average salary.

But the State pays this salary to each of its highly selected servants whether he is a bachelor, or married and has a small or large number of children. I have no hesitation in saying that this is an anachronism which constitutes a grave injustice to those civil servants who undertake the responsibility and labour of rearing families. The practice is a survival from the times when marriage and the production of a normally large family was more nearly universal than it is at present. There can be no doubt also that these civil servants represent just that class of men in which the tendency to postpone marriage and greatly to restrict the size of the family is especially strong and steadily growing. Many factors co-operate to bring about this tendency; but the comparatively small income is undoubtedly the most important and underlies most of the others.

The average income of £700 a year is not attained by the civil servant under the present system until he has been some years at work; and, as he does not enter the service until he is about twenty-four years of age, he will, on the average, not attain it until he is over thirty, perhaps forty, years of age. Now, to a highly educated man of cultivated tastes, one who has been brought up in the enjoyment of every kind of physical and mental luxury of the better sort—travel, sport, art, abundance of books and tasteful surroundings—an income of £700 a year seems barely sufficient to maintain in decent comfort a wife and one or two children, and to give to those children all those advantages of education which he himself has enjoyed; especially when, as is the case with the class we are considering, he is compelled to live in London, and therefore must afford his family long and expensive holidays in the country, or be content, like so many others, to see his children growing up physically degenerate. For there is deeply implanted in the breast of every Englishman of the better sort the healthy and admirable feeling that he is bound to afford his children opportunities of education at least equal to those which he himself has enjoyed. If then such a man marries, he marries late, and he almost inevitably restricts the size of his family to one, two, or, at most, three children, or perhaps abstains from parenthood altogether. And the temptation to remain a bachelor may be strong; for with £700 a year a bachelor can procure for himself all the luxuries that any intelligent person need desire; whereas the support of even a very small family demands strict economy and the sacrifice of many of the lesser enjoyments of life. No bachelor, in fact, needs more than £500 a year for his personal comfort and luxury, and with that income he may be regarded as at least equally well-to-do with a similar man who draws a salary of £1,000 a year and has to support a wife and five children.

That the man who in these circumstances marries and brings up a family of several children is performing a service to the State is as indisputable as the fact that the bachelor shirks this primary duty of the citizen. It is clear, then, that in the case of this class of civil servants the present system of remuneration not only constitutes a grave injustice to many of them, but leads directly to a great restriction of the numbers of the offspring of these servants, and therefore restricts artificially the production of those individuals of eminent abilities whose value to the State is incalculably great. Justice and expediency alike call urgently for a reform of the system. I do not think that any objection of appreciable weight can be made to such reform; it should therefore be welcomed, if only as an act of justice, even by those who may believe that no great advantages will accrue from it.

The reformed system of remuneration of these civil servants would consist in the adjustment of their salaries according to the number of their children.

If the first-class clerks of the home civil service were the only class of persons to which this scheme is applicable, its adoption might be regarded as a matter of relatively small importance. But I have dwelt upon their case merely because they exemplify the principles concerned in the clearest possible manner. There are numerous other posts in the pay of the home administration that are filled by highly selected men—inspectorships of education, of factories and so forth—to all of which the scheme is applicable. Again, the Indian and Colonial civil servants are a class selected by almost equally severe tests; and it may be hoped that the present agitation for reform in the Army may in the near future bring it about that the same will be true for the officers of both the Navy and the Army. The State will then have in its pay many thousands of public servants chosen from the very flower of the youth of the whole country on the ground of mental ability and physical and moral soundness. The application of the reformed method of remuneration will then assume an importance of the first magnitude, an importance that cannot but increase; for, however much we may regret it, there exists an unmistakable and apparently inevitable tendency for the State to assume wider functions, and therefore to employ larger numbers of servants of the highest ability procurable.

The inferior posts of the civil service are also filled by persons chosen by a process of extremely keen competition, and there is good reason to apply the scheme to their remuneration also, although in their case the matter is of less importance.

But we may go further. The adoption by the State of some such scheme as that suggested might have far-reaching consequences in inducing other institutions to adopt a similarly just and nationally beneficial system of remuneration. The example of the State might well be followed by the universities. At the present time the occupants of the university chairs receive stipends ranging from about £200 to £1,000 a year, except in a very few cases of larger salaries, and the average stipend is probably about £600 a year. These professors are selected for their exceptional abilities by a process of competition even more severe and more wide-reaching than that by which the first-class civil servants are chosen, and they form a group of which the average ability is probably distinctly higher, including as it always does a number of men of the very highest intellectual distinction. Our universities are growing rapidly in both size and number, and the number of professors, already very considerable, is rapidly augmenting. It is therefore a matter of extreme importance for the universities themselves, and for learning and research, that the occupant of a chair should be remunerated in such a way as will enable him to enjoy a normally comfortable domestic life, else the best intellects will not be drawn into the profession in sufficient numbers; and from the point of view of the future generations it is still more important that for such men the bringing up of reasonably large families should not entail hardships, the prospect of which must tend to restrict very largely the number of their offspring.

There is good reason, too, for the application of the proposed system to the remuneration of the selected servants of our organs of local government, the scope and importance of which are continually growing. And, if the example were set by all these public institutions, it may reasonably be hoped that the justice and the expediency of the system would be universally recognized, and that it would be adopted by all persons and institutions—especially such semi-public bodies as the great railway and steamship companies—that employ servants selected from the great average mass on account of their superior abilities and moral character.

If this happy result should ensue, we might confidently expect great benefits to the national breed and to the national power and capacity for progress—benefits compared with which any results that could be achieved by the elimination of the unfit by the most ruthless and autocratic of governments would be insignificantly small.

On the other hand, it is but too certain that, if some such system is not extensively adopted, there will be manifested in an ever-increasing degree the tendency, already deplorably real, for the most desirable elements of the population, moved by prudential considerations, to restrict the number of their offspring. This tendency has not yet been at work for many generations; it is the result of the increased strenuousness and the increased severity of competition of modern times. But there can be no doubt that, if not counteracted in some way, it will grow stronger, and that in the course of a few more generations it must, even if it should grow no stronger than it is at present, produce a most serious deterioration of the national breed. See what must happen. The upper strata, being relatively infertile, must be continually recruited from below; and this process, continually draining the great mediocre mass of its best elements, must result in a lowering of the average civic worth of that mass, so that in course of time the best that it can provide for the recruiting of the upper strata will be of less and less worth from generation to generation: and the more perfect our social ladder the more rapidly must this process of exhaustion go on. Our present social system is, then, one which tends in an ever-increasing degree to eliminate the most fit and the most desirable elements of the population, to select them by an elaborate organization from all the strata of society, to sterilize them and to replace them in the succeeding generations by inferior elements. No people, whatever its vigour, can fail to deteriorate under such a system; in the great world-struggle it must inevitably succumb and fall back to an inferior position among the nations. No improvements of institutions, of education, of environment can compensate for this. To those who, like the present writer, believe that the English-speaking peoples have evolved through long ages of strife and toil a system of civilization higher than any other that the world has yet seen, one full of promise for the future welfare and progress of mankind, the change of custom here suggested will not seem to be a matter for academic discussion merely, but will appear rather as the most urgently needed social reform that lies within our power to effect.

Increase of knowledge is showing us that we are responsible for the welfare of the generations to come in a fuller sense than could be realized by our fathers. We cannot shrink from the burden laid upon us and retain our self-respect.

X
FAMILY ALLOWANCES AS A EUGENIC MEASURE

IN the year 1906 shortly after the foundation of the Sociological Society of London, I opened a discussion at one of its meetings by advocating the institution of family allowances as a eugenic measure.[1] Although a number of eminent persons took part in the discussion (including Francis Galton, the initiator of the modern eugenic movement) I failed to elicit any support for the proposal, and my later efforts in the same cause seemed to have been equally futile.[2] The following considerations make the present moment seem opportune for a new statement of the case.

In 1906 I was able to point to only one institution in which the payment of family allowances was a well-established practice, namely, the Wesleyan Church, which had long ago adopted the practice of adjusting the pay of its missionaries to the needs of their families. In the interval the practice has been very widely adopted in several European countries and in Australia. Varied motives have brought about this new state of affairs; but, except in one case presently to be mentioned, the eugenic motive seems to have played no part, or but a very minor one.

In France, where the payment of family allowances to wage-earners was initiated soon after the War and has become the rule throughout many of the chief industries, the predominant motive seems to have been the desire to

[1] My remarks together with the discussion which they provoked were published in the second of two volumes of *Sociological Papers*, London, 1906, under the title, 'A Practicable Eugenic Suggestion'.

[2] Cf. my *National Welfare and National Decay*, 1921, and my *Ethics and Some Modern World-problems*, 1924.

promote the general birth-rate, that is to say, crudely put, the desire for larger supplies of cannon-fodder.

In Australia, where the introduction and wide adoption of the practice was largely due to the enlightened efforts of Mr. A. B. Piddington, chairman of a royal commission on the basic wage, the desire to promote the general birth-rate, already very low, in order to fill the vast vacant spaces which invite the invasion of Asiatic immigrants, was reinforced by the purely economic argument. Mr. Piddington seems to have been the first to recognize the absurdity, the injustice, and the economic waste involved in the prevailing system, based, as it was, on the assumption that every adult wage-earner is responsible for the support of a wife and just three children, no more and no fewer. 'I have shown,' he wrote of Australia, 'that this amounts to postulating the existence of 2,100,000 workers' children whom we know to be non-existent.' And 'the last six months have teemed with manifestations of the mischiefs resulting from the present chaotic condition of basic wage law. There is only one way out of these mischiefs, viz. child endowment on a national scale.' [1] It seems that in New South Wales, and in New Zealand also, family allowances are now paid by the State to all lower-paid wage-earners.

In Great Britain during the War non-commissioned officers and privates on active service were paid family allowances proportioned to the number of their dependants. Commissioned officers were denied this measure of elementary justice; and thus, since the whole population of the Empire was finely combed for men fit to play the role of commissioned officers, what might have been a great eugenic measure was made into one of opposite tendency. The principle adopted was: pay your lower ranks an adequate family allowance and promote their birth-rate; deny this benefit to your higher ranks, your class carefully selected from the whole population on grounds of vigour,

[1] *The Next Step, A Family Basic Income*, Melbourne, 1921.

efficiency, trustworthiness, and intelligence, and thus depress their already lamentably low birth-rate. Since the War, a similar dysgenic system has been instituted for the civilian population of Great Britain: family allowances for the unemployed and the unemployable are provided by the national government.[1]

More recently Miss Eleanor Rathbone [2] has vigorously advocated a national system of family allowances and has created a strong society for the promotion of her crusade. In this case the purely economic motive and the desire for justice as between man and man[3] seem to have been strongly supported by the specifically feminist motive, the desire to better the position of women by paying the child-allowances directly to the mother.[4] At first this movement seems to have been indifferent to the eugenic and dysgenic possibilities of family allowances. But I am glad to find in conversation with Miss Rathbone that this is no longer the case.

Lastly, it seems that in four dioceses of the Church of England the admirable example of the Wesleyan Church has recently been followed.

Other European countries, notably Belgium, have followed the lead of France, and have even gone beyond her in enthusiastic application of the family allowance principle to wage-payments in all the major industries.[5]

I have knowledge of only one instance of adoption of the

[1] It is worth noting that the justice of the principle receives some slight recognition in the reductions of income tax allowed on grounds of family responsibilities. But the reductions allowed are still so small as to be of negligible effect.

[2] Member of Parliament for the provincial universities and author of *The Disinherited Family*, London, 1920.

[3] She points out that in British industry only 9 per cent of the wage-earners have just three dependent children, while 52 per cent have none.

[4] In accordance with the principle of 'Endowment of Motherhood', Mr. H. G. Wells, with his genius for suggesting the wrong thing, seems to be the author of this slogan.

[5] I believe that they are paid also by several European States to certain classes of civil servants. But I have no definite information on this point.

system with some regard to its eugenic possibilities: namely, Sir William Beveridge, the head of the London School of Economics, has had the courage and enlightened public spirit to introduce the system in the payment of his distinguished staff of teachers. He tells me that he is well pleased with the working of the system, that he believes his staff also are in the main equally pleased; and he estimates that, although the system has been at work some few years only, it probably may claim to its credit some forty little 'economists'.[1]

It is clear, then, that while the justice and the economic expediency of family allowances are now widely recognized, their power for good or evil over the composition of any population has as yet received only very slight practical recognition. But a conjunction of other circumstances emphasizes the need and the opportunity for such recognition.

In all countries the economic depression is forcing men to seek new and better practices, juster systems of remuneration, wiser systems of distribution of spending power. We are being compelled to devise on the largest scale schemes of national economy. And especially is this true of the United States of America, where high wages and the general 'relative' prosperity prevailing up to 1929 had prevented any widespread interest in family allowances.[2] The period of recovery from the present economic depression will offer a most favourable opportunity for the introduction and spread of the family allowance system, for it will be a period, not only of general readjustment, but also one of rising wages and salaries.

Again, it is beginning to be widely recognized that the

[1] It seems that an Educational Allowance Fund was approved by the Governors of the School in 1925. From this fund £30 is to be paid annually to each member of the regular staff on account of each child under thirteen years of age and £60 for each child between thirteen and twenty-three years who 'is in full-time attendance at an approved place of education'.

[2] Only one American publication on the topic is known to me, a book by Prof. Paul Douglas, *Family Allowances*, New York, 1930.

policy of a State, the laws and customs of a nation, may profoundly affect the composition and quality of the population for good or ill. The restriction of immigration into America is perhaps the largest measure of such practical recognition hitherto achieved; and it is not putting it too strongly to say that this measure was instituted none too soon if American civilization was to have any hope of surviving its present disorders.

More recently the National Socialist Government in Germany has evinced a lively interest in the quality and composition of the nation; and the announcement of its intentions, whatever view we take of them, is rendering the service of compelling world-wide attention to the problems of eugenics.[1]

In addition to these two outstanding instances of national action prompted by concern for the quality of the population, the eugenic movement has made progress in many countries; and, especially, the argument for *negative* eugenic measures (sterilization of the feeble-minded, etc., etc.) has secured a wide hearing and led to some actual practice.[2] But negative eugenic measures can bring only very limited benefits; and the present state of the world is preparing men's minds to recognize that *positive* eugenic measures are the most urgent need of our time. The fact is forced upon us that we are no longer equal to the complexities of our civilization; that these complexities grow greater with every new application of the discoveries made by the physical sciences, that they, therefore, make ever greater demands upon the intellectual and moral qualities of our population; that we require for our guidance amidst vast

[1] 'The compulsory sterilization of those considered in the light of medical science . . . to be by heredity unfit has been introduced into Germany in a Bill "for the avoidance of inherited disease in posterity", which has been approved by the Government. The Bill is a logical product of Nazi ideology, with its conception of the community of race and blood and of the reaction of a healthy national organization against foreign bodies within it.'—*The London Times*, July 24, 1933.
[2] Notably in California.

national and international problems an increasing supply of men of the finest moral fibre and largest intellectual grasp; and that the supply is far from equal to the need; that we are short in leadership in every walk of life; that the old belief in the unlimited national reserves of potentially great men was founded merely in our will-to-believe, and is woefully contrary to the fact.

Two features of the present state of the world especially bring home to us our need of positive eugenic measures. First, all nations throb with the desire for peace and the horror of war: yet every nation lives under the menace of new wars and feverishly piles up fresh armaments; and all the multitude of organized efforts to ensure peace seems to achieve nothing. Secondly, in the midst of means of production of all material commodities so greatly improved that all men might live in plenty, many millions are racked by poverty and very few escape great economic anxieties. In satisfaction of the urgent need for leaders in every walk of life, the so widely urged negative eugenic measures, even of the most thorough kind, have little to promise. At best they can but relieve us of a special part of the burdens that weigh upon us, burdens which, if we were but sufficiently intelligent to order our societies effectively, would be but trifles.

In England a number of men of science have recently published articles strongly endorsing my proposal of 1906 and urging the institution of family allowances as a measure of great eugenic possibilities.[1]

Lastly, at the time my suggestion was put forward the evidence on which I had to rely in forming the premises of the argument was of a general and inexact nature. In the years that have elapsed a multitude of statistical and psychological researches have greatly strengthened these

[1] The most weighty of these utterances is that of Professor R. A. Fisher, the eminent mathematical biologist, in an article in the *Eugenics Review* and in the *Family Endowment Chronicle*, November 1931, entitled 'The Biological Effects of Family Allowances'.

premises, rendering them relatively exact and quantitative. I propose to state very concisely these premises and to summarize this new evidence.

(1) In the nations of our Western civilization (I write with special reference to Great Britain and America) the operation of 'the social ladder' effects a sifting of the population into social strata; the more capable persons, those better endowed physically, morally, and intellectually, tend to rise in this social scale of strata, or, if born in the upper strata, to maintain themselves therein; the less capable tend to sink or to remain at or near the bottom of the scale. This is true in spite of all the accidents of health and fortune, in spite of the inheritance of wealth and the handicaps of poverty; indeed, it is often asserted, not altogether groundlessly, that the inheritance of wealth is a positive detriment to the child of wealthy parents, while poverty in youth stimulates to effort and self-development and achievement. The social ladder has operated in this way for many generations; and with the advance of democratic freedom, the increasing opportunities of education for all, the increasing effectiveness of methods of selection and promotion in many callings, and an increasing popular prejudice in favour of the man of lowly origin,[1] it may be said to have attained at the present time to a very high efficiency. The results are: first, that at the present time a very much larger proportion of the naturally gifted are to be found in the upper social strata than in the lower; secondly, that naturally gifted children are procreated in very much larger proportion by the upper than by the lower social strata. This premise of my

[1] In America this prejudice is very distinctively a factor of importance. For example, a man who, like Herbert Hoover, was born and bred a farmer's son or one who, like Al. Smith, was bred on 'the sidewalks of New York' carries through life a distinct advantage as a candidate for office of any sort, as compared with a man like President Roosevelt who was born to an inheritance of wealth and culture.

argument was in 1906 an inference from a multitude of biological and social facts, an inference which, though it could be made with a degree of probability amounting almost to certainty, was nevertheless so repugnant to our more generous sentiments that very few could be induced to accept the conclusion. To-day it is a fact well-established by the convergence of many independent researches. Since the conclusion is the principal basis for any large-scale eugenic policy, it is worth while to summarize very briefly some of the more important evidences.

One of the earliest exact researches, though on a small scale, was made under my supervision and suggestion by Professor Cyril Burt, then a graduate student at Oxford. He applied, individually and with great care to exclude disturbing factors, a series of specially designed mental tests to the boys of three schools: (a) a private preparatory school attended mainly by the sons of professional men; (b) a rate-supported school of the best class attended mainly by sons of well-to-do tradesmen; (c) a rate-supported school attended mainly by sons of labouring men. The differences of level of achievement were very marked and in the order a, b, c.

Professor L. M. Terman of Stanford University has made the study of gifted children his life work.[1] By the application of standardized mental tests to very large numbers of children in the schools of California, he has found a number of children whose performance under such testing greatly excels that of the average child. The average child (an average based on very large numbers tested) achieves in a degree expressed by the intelligence quotient (I. Q.) 100. Selecting from all the schools of the State all those children whose performance is expressed as I. Q. 130 or better (they range up to or near to I. Q. 200), Terman has made these 'gifted children' the objects of intensive study. He finds that among the fathers of these 'gifted children' fifty-three per cent are of the professional class; thirty-seven

[1] cf. *Genetic Studies of Genius*, University of California Press, 1925.

per cent are clerical workers; ten per cent are skilled manual workers; while among the semi-skilled and unskilled manual workers such fathers and such children occur in negligible proportion only. His most striking general conclusion is that the professional class, which constitutes about two per cent of the whole population of the State, produces more than fifty per cent of the children of high natural endowments.

Closely similar findings by aid of similar methods have been made independently in the schools of New York City, and again in those of Madison, Wisconsin; and, by Dr. Godfrey Thomson, in those of the English county of Northumberland. The Wisconsin observations are interestingly presented as follows: Children of the professional class achieve a mean I. Q. 115; those of clerical workers have mean I. Q. 106; those of business men, 104; those of skilled labourers, 99, of semi-skilled, 92, and of the unskilled, 89.

Several observers have made comparisons of the kind first made by Burt, comparing by the aid of mental tests children of schools of distinct social classes in England and America; and they have found closely similar results.[1]

But the layman is still unconvinced. He is apt to entertain the popular belief that the clever child is a bespectacled monstrosity of poor physique, of unstable neurotic disposition and shady character; and, being oftener than not an incurable sentimentalist, he says: Give me for building a nation the children of the brawny sons of toil, bronzed with health and safely sane from labour of the hands in field and factory.[2]

[1] e.g., in the Horace Mann School and the Ethical Culture School of New York.

[2] Professor Leta Hollingworth, who in her volume *Gifted Children*, New York, 1926, reviews and judiciously summarizes all the evidence available at that date, writes: 'The manual labourers outnumber the nobility in England one hundred times, but produce only one quarter as many children who achieve eminence. This and many other similar facts are contrary to popular belief. The very exceptionality of the rise of a man to extreme eminence from the humblest

But here again we now have the results of careful large-scale research to go upon. Professor Terman finds that his 'gifted children' are well above the average in height and weight, in health, in quickness and efficiency of movement, and also in trustworthiness, power of attention, persistence, unselfishness, courage, self-control, initiative, leadership. 'Children selected wholly by intelligence tests . . . show desirable traits of character and temperament in superior degree.' Further, it has been shown by various workers that in the main the position of children in the I. Q. scale remains unchanged throughout the years of schooling; and that, in respect of achievement in the mental tests, the correlation between twins=·80, that between siblings=·50, that between cousins=·25, while, of course, the correlation between haphazardly chosen groups is zero. All of which goes to confirm the view that achievement of this sort depends largely upon inherited qualities (though not, of course, wholly), that superior achievement is largely the expression of superior heredity.

These results of experimental research, remarkably concordant with one another and with all well-founded expectation, do not stand alone. The conclusion to which they compel us is supported by a number of statistical studies of another kind. Dr. McKeen Cattell has studied the parentage of Americans distinguished in the field of science.[1] He finds: 'The professional classes have contributed in proportion to their numbers about fourteen times as many scientific men as the others; the agricultural classes only half as many as the manufacturing and trading classes. The farm not only produces relatively fewer

rank of life is sufficient to fix it in the public attention, so that it is remembered. In this way develops an illusion that most eminent men have been poor in their youth.' Confronted with the thesis here presented, the American layman almost automatically refers to Abraham Lincoln, while the Englishman similarly refutes it by references to George Stephenson or James Watt and to countless mute inglorious and purely hypothetical Miltons.

[1] *American Men of Science*, New York, 1921.

scientific men, but a smaller proportion of them are of high distinction.'

Dr. S. S. Visher[1] has studied the origins of the persons who appear in the American *Who's Who*, a population which may fairly be regarded as well above the average in achievement and in the qualities that make for achievement, and as comprising a very large proportion of all the most highly talented members of the nation. Among his findings are that †he cities produce six times as many such persons as the farms; that, of the fathers of this group, 70 per cent were of the professional and business classes, 25·4 per cent were farmers, 6·3 per cent were skilled or semi-skilled manual workers, ·4 per cent were unskilled labourers. Or, putting these conclusions in another form, if the value of an unskilled labourer as a potential producer of children who will achieve a place in *Who's Who* is stated as one, then the corresponding value of a man of the skilled and semi-skilled classes is thirty, that of the farmer is seventy, that of the business man is 600, and that of the professional man is 1,400.

Similar studies have shown that in the past, both in New England and in Old England, the clergy have been a prolific source of men and women of high achievement. And the Scottish manse is notorious as the source of much ability of varied kinds.

Studies of a third class converge to the same conclusion: Dr. F. A. Wood's study of *Mental and Moral Heredity in Royalty*,[2] and studies of gifted men and families by Francis Galton, Karl Pearson, Havelock Ellis, A. T. Gunn, and others, have shown unmistakably how exceptional vigour and general ability, as well as special forms of talent, are propagated in families, often through many generations, and that, both in England and America, a very large proportion of the most distinguished and effective persons of the nation have been produced by a few families or stocks, e.g. the stock of the Randolphs, of Jonathan Edwardes, of

[1] *American Journal of Sociology*, 1925.   [2] New York, 1906.

John Adams, in America; the Churchill, the Darwin-Galton, the Haldane, the Wordsworth, the Arnold stocks in Great Britain.

Terman's showing cannot be escaped; it is established, at least for Great Britain and America: 'The occupational group which is least numerous, the professional, furnishes by far the greatest proportion of gifted offspring, and the most numerous groups of the population furnish very few.' It only remains to follow up the 'gifted children', to find what degrees of distinction and achievement they will show in adult life and what proportion of 'gifted children' they will produce. But only a perfectly irrational person can doubt the general nature of this future harvest. Terman has already shown that a group of the parents of 'gifted children' manifest an I. Q. far above the average.

(2) The second main premise of the argument is that the individuals and social classes which are potentially the most fertile in children of talent and distinguished achievement are at present, and for some generations have been increasingly, of low birth-rate.

In support of this premise abundant evidence has accumulated. In 1906 I was able to cite Professor Karl Pearson to the effect that 'Statistics are forthcoming, and will be shortly published, to show that the families of the intellectual classes are smaller now, very sensibly smaller, than they were in the same classes fifty years ago; that the same statement is true of the abler and more capable working and artisan classes; but that as you go down in the social grade the reduction in size of families is less marked'.[1] Such statistics have now been published in abundance, notably by Dr. David Heron for the population of London, and by Dr. Louis Dublin for America, and by Professor Hermann Muckermann for Germany. It has been shown that birth-control (including under this term late marriage and various other forms of restraint, as well

[1] *National Life from the Standpoint of Science*, London, 1905.

as the use of contraceptive measures on an ever-increasing scale) is rapidly bringing the populations of England and America to a stationary state; that if these populations do not actually decline in numbers it is only because those strata which (as shown above) produce a negligible proportion of gifted children continue to multiply, while the higher social strata are maintained only by constant recruitment from below—a drainage process which can but impoverish still further the strata from which the recruits are drawn.

Studies of the graduates of various universities have shown that these persons, men and women alike, do not produce enough children to fill their places. They are the cream of the whole population; and the cream is skimmed off from each generation and thrown away. Birth-control has, in short, given enormous scope for differential fertility to work changes in the composition of these populations; and it is working most disastrously.

In the small space at my disposal the facts can best be brought home to the reader by citing the conclusions of some authoritative students. A distinguished American biologist who has devoted much study to problems of population, after pointing out that, in America, 'over extensive areas the native population of native parentage is no longer reproducing itself and is gradually being supplanted by more prolific alien peoples', writes of England: 'From what is known of fertility in relation to occupation and social status we may count among the disappearing stocks the professional classes, the more successful business men, the skilled artisans, and the bulk of the more educated and thrifty classes. Only those strata which are least affected by the decline of the birth-rate are continuing to reproduce themselves. There is no way of avoiding the conclusion that *the country is suffering a frightful loss* of its best hereditary strains. Unless the relatively high fecundity of subnormal and dull normal humanity can somehow be checked and unless *the relatively low fecundity of the well endowed can somehow be increased, racial deterioration seems inevitable.'*[1]

And in another article the same authority writes: 'The misuse of birth-control has done a tremendous amount of damage and promises to do more. . . . The problems created by the growing practice of birth-control are far more serious than is commonly realized. They will not be solved by the easy method of *laissez-faire*, nor by fulmination against the wickedness of contraception. One of the greatest problems facing civilized mankind is how to secure the undeniable economic, humanitarian, and eugenic benefits of birth-control and at the same time escape from its very real dangers.'[2] And in a third article: 'Some day the people of Great Britain may be stirred to do something really effective towards reducing their burden of defective humanity, but at present they seem to be doing everything they can to increase it.'[3]

The most eminent and authoritative of biological statisticians writes: 'It cannot be doubted that the more capable and energetic men do, on the average, attain to higher skill and secure higher wages; and our educational system is largely concerned in drafting the abler children of the poor into better paid occupations. The inference is obvious that these occupations contain more than their share of innate ability, while they contribute less than their share to the next generation. That the better paid should be replaced wholesale, at the present alarming rate, by the children of the less successful, can only mean the biological elimination from the race of the qualities which make for successful citizenship. As a matter of calculation this elimination must be now proceeding so rapidly as to produce very serious evolutionary consequences in only a few generations'.[4]

[1] Professor S. J. Holmes, 'Is England approaching Depopulation?' in *Journal of Heredity*, March 1931. Italics are mine.
[2] *The Scientific Monthly*, March 1932.
[3] *Journal of Heredity*, February 1932.
[4] Dr. R. A. Fisher, *Family Endowment Chronicle*, November 1931.

Again in a recent article Messrs. E. M. Hubback and M. E. Green write: 'Mcdougall's second contention— that the higher grades of society were contributing much less than their share to the growth of the population— received startling confirmation in Dr. Stevenson's analysis of the 1911 census; and by 1921 the birth-rate of the upper and professional classes had fallen to nearly half that recorded for unskilled labourers—98 per thousand married men, as compared with 178. Moreover, the statistics indicate not only a relative, but an absolute infertility. Dr. R. A. Fisher has pointed out that, for parents with £300 a year or over, the current supply of children is not sufficient to replace more than one-half of the parental generation.'[1]

Perhaps the most generally significant figures on the differential birth-rate for England and Wales are contained in the following table, which gives the number of births per 1,000 married men under fifty-five years of age for the year 1921.[2]

Upper and middle class	98
Intermediate class	105
Skilled workmen	134
Intermediate workmen	152
Unskilled workmen	178

The most thorough and convincing evidence of this kind results from Dr. H. Muckermann's statistical studies of the reproduction rates of the classes of certain typical German communities and of the academic population of all Germany. In the former study[3] he shows: 'In the country families [peasants] the number of families with 0–3 children

[1] *The Eugenics Review*, April 1933.
[2] These figures are taken from *Social Structure of England and Wales* by Drs. Carr-Saunders and C. Jones.
[3] 'Vergleichende Untersuchungen über differenzierte Fortpflanzung in einer Stadt- und Landbevölkerung (1,847 Familien mit 7,201 Kindern)', *Zeitschr. f. inductive Abstammungs- und Vererbungslehre*, Bd. LXII. The author cites other German studies of similar tendency; e.g., Dr. Kurz studying 18,735 families in Bremen shows that the families which send their children to the higher

is fifteen per cent of all, while the number of families with four or more children is eighty-five per cent. In the families of industrial labourers [*Arbeiter*] the corresponding figures are thirty-two and sixty-eight per cent; in those of craftsmen [*Handwerker*] thirty-six and sixty-four per cent; in those of shopkeepers forty-four and fifty-six per cent; in those of the lower officials the numbers are equal. Among the higher officials the figures are reversed, the families of three children or fewer are fifty-four per cent, and those of four or more are forty-six per cent. In the families of the academic group we have sixty-eight per cent of the marriages giving only three or fewer children, while only thirty-two per cent give four or more children.' Or, more concisely, the proportion of families with three or fewer to those with four or more children is among the peasants 0·16, among the intermediate strata taken together 0·58, and in the academic class 2·16.

An earlier study of 3,947 families of professors in the German universities and technical high schools by the same investigator shows that this class produces only 1·65 children per married couple, a number very far from sufficient to replace the parents. It shows also that this state of affairs was already pretty well established many years before the onset of the recent rapid decline of the the birth-rate of the German population as a whole. While the professor and his wife contribute 1·65 children to the make up of the rising generation, the peasant couple contributes 4·2; and the disparity is in no degree offset by infant mortality, which is surprisingly low in the peasant group studied.[1]

As regard the largely voluntary nature of the low fertility

schools have on the average less than two children; forty per cent of them have no more than one child, thirty-three per cent have two children, twenty-five per cent only have more than two children; that is to say, the social stratum which avails itself of the higher schools is constantly dying out.

[1] 'Differenzierte Fortpflanzung', *Archiv für Rassen- und Gesellschaftsbiologie*, Bd. **XXIV** (1930).

of the upper social strata there is no room for doubt; though, by the nature of the case, conclusive evidence on a large scale is very difficult to obtain. Dr. Cattell has made one such inquiry and finds that, among 461 men prominent in science, 285 acknowledge voluntary limitation of the family, and that six in seven men of the whole group desired to have no more than two children.

Dr. Muckermann's figures afford evidence of the same kind: thus while in the peasant families the first, second, and third five-year periods of marriage yield 2·3, 2, and 1·8 children respectively, in the same periods the academic marriages yield 1·65, 0·65, and 0·3 children; a suddenness of falling off in the latter class which points to voluntary control of some kind as a main factor.

In my discussion of 1906 I assumed that the infertility of the upper social strata was in very large measure voluntary and was largely founded in economic considerations, in motives not wholly selfish, not a mere desire for comfort, but a desire to give the children a good start in life, a desire to spare the wife the burdens and the dangers to health inseparable from the bearing of a large family.[1]

[1] My long residence in America among the well-nigh servantless families of university teachers has shown me that the economic motive plays there a great role. In the United States, where university salaries are hardly equal on the average to the wages of skilled artisans, the mere cost of bringing a child into the world is, under present social conditions, a very serious expense for the professor, one equal roughly to two months' salary. Here again Dr. Muckermann's figures afford interesting confirmation; namely, he finds that while the number of children per family of the university professors is lamentably low, the corresponding figure for the professors of the high schools is significantly lower. Now I have no figures that enable me to compare the salaries of the two classes, but there is good reason to believe that the professors in the universities receive on the average rather higher salaries than those in the high schools. Another feature making in the same direction it is perhaps worth while to point out. The University of Leipzig stands near the bottom of the list with 2·13 children per family and seems thus to illustrate the notorious influence of the *Gross-stadt-leben*, while the University of Berlin (where salaries are I believe considerably higher than elsewhere) stands near the head of the list with 2·98 children per family; yet Berlin is more *gross-städtlich* than Leipzig.

And in urging the addition of family allowances to the salaries of all selected classes from skilled artisans upwards, I dwelt only on the effect which such allowances must have in removing or weakening (according to their magnitude) the economic motive for late marriage and family restriction.

But the case for family allowances is even stronger than I had supposed; for (as Dr. R. A. Fisher has made clear) they may be expected to work very favourably in a manner quite independent of, and over and above their influence on, the economic motive to family restriction.[1] The infertility of the upper social strata (whether mental or physiological) is, Dr. Fisher argues, largely due to the fact that advancement in the social scale is, and long has been, more easily effected by families and members of families of few children; he infers that the natural lack of fertility has thus been concentrated in ever-increasing degree in the upper social strata. Family allowances would put an end to the further advance of this process, one which already is accountable for the lamentable frequency of married couples of high distinction who remain childless in spite of the best intentions.[2]

In another way the possible scope and value of family allowances has been shown to be much greater than I claimed in 1906. At that time I could not see any hope for the extension of the system beyond the salaried classes working for the State and local governments, for universities and for other semi-public bodies such as the great railway companies. For in competitive industry it seemed inevitable that, if family allowances should be established (by law or otherwise), the bachelor and the man of very

[1] *The Genetical Theory of Natural Selections*, Oxford, 1930. In pointing out this mode of working of family allowances (which had totally escaped me) Dr. Fisher has rendered a great service. He regards it as more important than the influence on the economic motive for restriction.

[2] Of my personal knowledge I could make a considerable list of such cases, a list of couples from whom an array of boys and girls of the highest promise might confidently have been expected, if only they had been fertile.

few children would have a large advantage in the competition for employment. But this difficulty and this limitation have been overcome by the intelligence and patriotism of the French industrialists, who have devised and widely established the *caisses de compensation.* 'This meant,' writes Mr. H. H. R. Vibart, 'that employers within an industry or within a given district joined together in one association and each of them paid into this association or fund so much per person employed or so much per cent of their wage bills. Out of that sum allowances on an agreed scale are paid by the fund to the workers, in accordance with the size of their families and without regard to the firms to which they belonged. This simple arrangement, capable of infinite minor modifications, met with immediate and sustained success', and now has become 'something taken for granted as part of the ordinary framework of the national life'.[1]

In 1906 I urged that a wise system of family allowances might, by perpetuating and multiplying the best qualities of our populations, bring greater advantages than all the most extreme measures of negative eugenics, and that such a system is open to no serious objections and difficulties such as lie in the way of almost all the negative measures. To-day this two-fold proposition is no longer a speculative suggestion: it has become an established truth.

It is clear, then, that without any great change of our social system, but in very simple ways, family allowances can be made customary throughout a very large part of the whole population; and it is equally clear that, like birth-control, they are capable of working disastrously if unwisely applied,[2] or of becoming a great agency for

[1] From a brief account of the extremely rapid spread of the system through French and Belgian industry, in *Eugenics Review,* April 1933.

[2] As in Great Britain at the present time, where this application is restricted to the unemployed and where (as an editorial in the *Eugenics Review* remarks) 'even aments now realize that they improve their situation by leaving their parents' homes and marrying on the increased allowance they receive from the rates and taxes'.

the preservation of the best qualities of the population, the qualities now being so rapidly weeded out from the populations of Great Britain and the United States of America. A widespread system of family allowances may, in short, be highly dysgenic or powerfully eugenic: and the essential difference between a dysgenic and a eugenic system is that in the former the allowance is made according to a flat rate, the same for all; while in the eugenic system the amount of the allowance per child is proportional to the wages or salary of the parent. The point is too obvious to need elaboration.

It is, then, with the proviso that the principle of family allowances be applied with an eye to their biological effects that we must accept the recent emphatic endorsements uttered by several students of our population problems; as when the editor of the *Eugenics Review* urges 'the strongest of all arguments for extending the principles of family endowment to those normal citizens whose children are national assets';[1] when Messrs. Hubback and Green write: 'A scheme of family allowances which would, it seems, be the only means open to us to preserve the qualities of eugenic value in the salaried and professional classes. . . . We can surely ask the State to set an example to other employers in adopting a measure which should prove a valuable contribution both to the problem of maintaining the best qualities of the race and of providing a favourable environment for the rearing of its future citizens'; when Sir William Beveridge states that the institution of family allowances would be the greatest single step towards the abolition of poverty that can now be made; and when Professor R. A. Fisher writes: 'even had they [family allowances] been economically disadvantageous, and therefore have involved some sacrifice of material well-being, it would still have been necessary to advocate their introduction as the most powerful available means of preserving among civilized

[1] April 1933.

peoples those innate qualities which make civilization possible'.[1]

Family allowances are so obviously just, so economically expedient, so politically advantageous, so powerful to promote the aims of the humanitarian and the feminist, that in a world racked with economic distresses and discontents, a world of rapidly falling birth-rates, a world deeply concerned to effect radical changes in its economic system, we may confidently expect to see them universally instituted in one form or another in the immediate future. Just as within a decade they have from small beginnings become basic features of the national economy of France, of Belgium, and of other European countries, so within the next decade they will surely become widely established in Great Britain and America, no matter how little or how greatly the forms of political institutions may be changed. And the great question is: Will they be given the dysgenic or the eugenic shape? If the former, it matters little whether capitalism, socialism, communism, or fascism shall prevail: for we shall all go down together in ignominious chaos. If the latter, then again it matters little what political forms shall survive the present struggle; for, if men are but sufficiently well bred, their societies will thrive under any political forms; and indeed, in the extreme case, governments will become superfluous luxuries, dwindling survivals from the present dark age of prejudice and superstition.

The danger that in Great Britain and America family allowances will be given the dysgenic form is very great. Can it be hoped that a democracy shall accept and everywhere practise an essentially aristocratic principle? Even so enlightened a pioneer of family allowances as Sir William Beveridge has introduced the flat rate, the dysgenic form

[1] loc. cit. Since this article was written yet another distinguished biologist has proclaimed the eugenic value of family allowances, namely, Dr. C. C. Hurst (author of *The Mechanism of Creative Evolution*) in concluding a paper on 'The Genetics of Intellect' read before the British Association in September 1933.

of family allowance, in his great school.[1] And if during twenty-seven years half a dozen biologists have been led to recognize the potential eugenic power of family allowances, how many decades will be required for the conversion of a majority of the now all-powerful economists? No, if our civilization is doomed to sink into ever deeper decay, is it not just because our society has already sunk below the critical level, the level of intelligence at or above which it might still achieve recognition of its greatest need, its main danger, its gravest malady, and its possible remedy?

[1] He is already realizing that this is, from the merely economic point of view, inexpedient. Where salaries range from £200 to £1,200 a year, a flat rate which nicely fulfils its functions at one end of the scale is either wasteful or ineffective at the other end.

XI

WAS DARWIN WRONG?

RATHER more than a century ago the great French zoologist, J. B. P. Lamarck, propounded the first modern theory of organic evolution. He pointed to the fact that animals of many species adapt their modes of life and, consequently, many details of their structures, to particular circumstances, modifying them, more or less intelligently, by means of their efforts to cope with those circumstances. He supposed that modifications of structure and function thus acquired are in some degree transmitted from parent to offspring. He suggested, further, that such acquisition and transmission of modifications have been the main factors in bringing about the differentiation of species and the progressive evolution by which higher forms of life have proceeded from lower.

This Lamarckian theory of evolution excited no great interest until Charles Darwin's *Origin of Species* led, in the course of a few years, after its publication in 1859, almost the whole of the scientific world to belief in evolution. Then the theory of Lamarck became generally accepted; for Darwin, while putting forward his new theory of the origin of species by 'natural selection', accepted at the same time Lamarck's theory of evolution by acquisition and transmission of modifications. Indeed, when once evolution had been accepted, as an historical fact, the process of transmission assumed by Lamarck's theory seemed so natural and obvious that the theory became popular; and to this day many highly educated men who believe that they have a good general understanding of the evolutionary controversy imagine that what naturalists mean by the word 'heredity' is just such transmission of acquired modifications as Lamarck had postulated. In this belief they are seriously in error. In science the word 'heredity' points only to the general truth that offspring in the main resemble their parents, inheriting from them and from remoter ancestors their general features and functions—namely, all those characteristic of the species, as well as many of the more special features which distinguish one individual from another.

Darwin's view of the process of evolution, combining his own theory of natural selection with Lamarck's theory of the transmission of acquired modifications, was for a time generally accepted by the scientific world. But in the eighties August Weissmann challenged the Lamarckian theory. He propounded the doctrine of the continuity of the germ-plasm and found in it strong grounds for doubting the validity of that theory. He taught that germ-plasm is an immortal substance which multiplies itself by growth and division; that the body of each animal is but a large lateral bud growing out from the continuing stream of germ-plasm; and that, although the body shelters and nourishes the germ-plasm, the latter is the sole bearer of the hereditary qualities, takes no part in the life of the body beyond assimilating from it food-substances for its growth, and is not affected in its composition or structure by the vicissitudes of the body, or not in any other way than in sharing, perhaps, in some degree the effects of good or bad nutrition. Weissmann was quickly followed by a large number of biologists who succeeded in casting doubt upon every alleged instance of Lamarckian transmission. Weissmann and his followers taught, then, that evolution has been the work of natural selection alone. This is the doctrine of neo-Darwinism.

Let me try, by the aid of an imaginary analogy, to make clear the differences between these three schools of thought—the Lamarckian, the Darwinian, and the neo-Darwinian. Suppose a feudal nation, the king of which awards to each of his principal followers within a conquered territory an estate whose boundaries are exactly defined. Each of these landlords transmits his estate to his heir. After many generations, it is found that the original estates have undergone considerable changes. Some have grown larger and more complex, and others have disappeared altogether from the map. There are two ways in which these changes may have been brought about.

First, a landlord may, by his own efforts, have annexed to his estate portions of adjoining territory; or, by failure to cultivate or in any way make use of parts of his estate, a landlord may have allowed his title to such parts to lapse, so that his estate has become smaller and simpler. It is obvious that this mode of alteration of estates can produce progressive changes of the boundaries from generation to generation only if the law of the land permits the landlord to transmit to his heir, in the modified form given it by his efforts or his neglect, the estate which he inherited.

Secondly, the Crown may have granted additions to some estates and have taken away parts of others; and in each such case the estate is transmitted in its changed form to later generations.

In this analogy, the estate stands for the constitution inherited by each individual animal; the Crown stands for nature—the laws of the land, or of the Crown, are the laws of nature. Additions to or subtractions from an estate resulting from the efforts or the neglect of the landlord are the modifications[1] insisted upon by Lamarck; additions and subtractions made by decree of the Crown are the spontaneously occurring variations which were made by Darwin the essential features of his theory.

Lamarck held that the law permits the transmission of an estate in the changed form given it by the efforts or the neglect of the landlord; and his theory was that all

[1] The term 'modifications' is generally used and is here used to denote only changes induced in the way to which Lamarck drew attention, namely, by the efforts of the individual creature.

the great changes found to have taken place in some estates after many generations have been produced in this way.

Darwin said that though it was true that estates changed gradually in the way Lamarck had supposed—namely, by transmission of acquired modifications—yet there was a mode of change overlooked by Lamarck; namely, changes (additions or subtractions which he called 'variations') are made by royal decree without reference or relation to the efforts or the neglect of the landlord. And he held that the great changes which had taken place in many estates were due largely to such arbitrary decrees of the Crown made from time to time.

The neo-Darwinians said that the law has always forbidden and made quite impossible any transmission of an estate in the changed form given it by the landlord. They maintained that an estate can be transmitted from father to son only as it was inherited by the father, together with any additions or subtractions which the Crown may have decreed to the father.

The reader may ask with surprise: 'But where, then, does "natural selection" come in? We had always heard that "natural selection" was the essence of Darwin's theory!' As a matter of fact natural selection is a secondary feature of Darwin's theory. Its primary and most essential assumptions are that many slight changes of estates are effected by decree of the Crown; that such changes are transmitted from father to son; and that, of the many great changes found to have occurred after many generations, many are the accumulated sum of such slight changes by decree.

The dispute between the neo-Darwinians and the Lamarckians, which has so fiercely raged and still rages, is then a dispute concerning the fundamental laws of inheritance. No one denies that changes of the kind postulated by Darwin (changes by decree—variations) occur and may be transmitted. And no one denies that changes of the other kind occur, that modifications are made by the efforts of individuals. The essential question in dispute is whether such modifications are transmitted from father to son, whether they are inherited in any degree, whether the laws of heredity permit such inheritance or transmission.

The critics of neo-Darwinism continue to assert the inadequacy of natural selection and the consequent need of some other principle such as that of Lamarck. They insist that the slight, unstable, fortuitous variations on which Darwin relied cannot have sufficient survival value to secure their own perpetuation. They point out that, even if some such slight variations of a favourable nature should have, in themselves, survival value, should here and there enable their possessor to survive where his fellows succumbed, yet this happy result would have to be repeated in innumerable instances if the new feature is to become established as a character of a species. And they urge that any such result is wildly improbable, because the favouring influence of any such favourable variation must in the vast majority of cases be offset and neutralized by coincident unfavourable variations: for, in the long run, of all purely fortuitous variations, the unfavourable must be far more numerous than the favourable.

There is much force in this last argument. The average animal is a marvellously balanced system of mutually adapted organs and interdependent functions; and any fortuitous changes in such a system must in the great majority of instances tend to disturb the nicety of its balance rather than to improve its efficiency. Imagine an immensely complex and delicately adjusted machine (for that is what, according to the neo-Darwinists, each organism is) and imagine that every now and then the fortuitous incidence of some force (from within or from without) displaces or changes some part of the mechanism. In what proportion of such instances can we suppose that the machine will be more efficient, in what proportion less efficient, by reason of such changes? Surely it would be a generous estimate if we assumed that one in a thousand might be favourable, ten neutral and harmless, and 989 unfavourable.

In order to meet this difficulty the neo-Darwinians have postulated variations of a kind not assumed by Darwin, variations of a larger, more complex, and more stable kind, which they call mutations. A mutation is conceived to consist in a group of mutually adjusted changes or variations; a group which is transmitted whole as a functional unit. If such mutations really occur as true novelties, one main difficulty of neo-Darwinism disappears; but it is not yet fully established that the alleged instances of mutation are more than pathological departures from the norm, due to some failure of the normal interplay of units of the germ-plasm. Further, if true adaptive favourable mutations occur with sufficient frequency to be the materials with which natural selection has effected all the steps of organic evolution, a new problem arises: Whence come these mutations? They cannot be regarded as merely fortuitous, as products of happy chance. In short, neo-Darwinism can be made to work only by postulating these mysterious favourable mutations—mysterious and utterly inexplicable unless they be the products of Lamarckian adaptation and transmission.

This controversy has interested a far larger circle than the biologists. The issue is of the deepest significance to all the social sciences, to philosophy itself, and to every man who takes an interest in the larger aspects of modern science and speculation. For the question in dispute not only lies at the very heart of the problem of evolution— the problem of how biological evolution goes on—but also it is at the heart of the greater issue between, on the one hand, scientific materialism and, on the other hand, every form of science, philosophy, and religion that would regard mind as in any sense real and operative in the universe.

Darwin's theory provoked world-wide interest and controversy because, for the first time, it seemed to render possible the explanation of the origin of all the multitude of beautifully adapted living organisms by a purely mechanical theory. For it was widely held that all those variations which we have likened to changes of estates made by royal decree are in reality undesigned and purely fortuitous, the chance or accidental products of the mechanical shuffling of inert counters, molecules, atoms, or what not. Yet Darwin himself had not accepted this purely mechanical theory of evolution; for so long as the Lamarckian principle of transmission of modifications was accepted (as it was by Darwin) it followed that, although it was no longer necessary to postulate a designing Creator of all living things, mind had nevertheless played a dominant role in biological evolution—mind immanent within each organism and working within it. For the efforts of each organism which produce its adaptive modifications seem to be expressions of mind, however lowly and obscure the purpose and the intelligence displayed. And the variations or mutations through which 'natural selection' was held to work in the manner of a pruning knife might be regarded, not as the products of fortuitous mechanical shufflings or of the arbitrary decrees of the great god Chance, but rather, in the main, as the products of intelligent efforts at adaptation to changing environment.

But when the neo-Darwinians denied the transmission of these products of mental activity, the modifications produced by intelligent purposive efforts, there seemed to remain in the drama of evolution no role for purpose and intelligence, no way in which mind might have guided its course. The denial seemed to forbid us to believe that purpose and intelligence had played any part, however slight, in producing, or in steering the course of, evolution. It seemed that if this denial were well founded, we should have to regard the whole of biological evolution as a purely mechanical or mindless process. Hence, in addition to the biologists, thinkers and writers of all types have taken a hand in the controversy over Lamarckian transmission. Herbert Spencer, Samuel Butler, Bergson, Bernard Shaw, and many others have used their wit and utmost dialectical skill in the effort to show that the neo-Darwinian theory will not work, and that it is necessary to supplement the theory of natural selection by the acceptance of the Lamarckian principle. But on the whole their efforts have availed little. The neo-Darwinians have been forced to concede that the principle of fortuitous variation is not a sufficient basis for the theory of evolution by natural selection alone; but they have continued to deny the possibility of any transmission of modifications.

The denial of the transmission of modifications has been very generally accepted by the biologists of Germany, America, and Great Britain. In France, on the other hand, a considerable number, probably the majority, of biologists have continued to believe in the reality and the important role in evolution of the transmission of modifications: an interesting illustration of the influence of non-rational factors in purely scientific theory: for Lamarck was a Frenchman.

'But,' the reader may exclaim, 'this is not a question to be settled by dialectics and a display of wit! It is, surely, a question for experimental decision! What have the experimental biologists been doing in their laboratories and breeding stations?' The answer is that many experiments have been made with a view to testing the Lamarckian principle; but, though some of them have brought results that seem to favour the Lamarckian theory, none have seemed conclusive, none have seriously shaken the confidence of the neo-Darwinians. Many of these experiments were not well designed with a view to give scope for the play of the true Lamarckian factor, namely, intelligent effort on the part of the creature; and that, perhaps, has been a principal ground of their failure to produce conclusive positive evidence.

I have long held that we should go boldly back to Lamarck and assume with him that the essential factor to be investigated is the effort, the more or less intelligent striving, of the organism to adapt itself to new conditions; and in the year 1920 I instituted an experiment designed in accordance with this specification. When my experiment had been running only some two years, the biological world was startled by a brief announcement from the Russian physiologist, Professor Ivan Pavlov, famous for his many delicate experimental observations on the nervous reactions of animals. Pavlov announced that, by training white mice of five successive generations to respond to the sound of a bell by seeking their food, he had been able to observe very clearly marked Lamarckian transmission of the modification of response thus induced. If these procedures and observations were above reproach, the long-debated question was finally settled in the sense favourable to Lamarck. But the intensity of the effect reported by Pavlov was beyond all reasonable expectation; and there were other reasons of a technical nature for hesitating to accept his results. I therefore decided to prolong my own experiment; and this decision was fortunate, for early in the year 1926, Pavlov, with the courage and honesty to be expected of him, announced that there was some fatal flaw in his procedure and retracted his claim to have demonstrated Lamarckian transmission.

My experiment consists in taking a stock of pure-bred white rats; dividing it into two equivalent halves; training each half-stock in successive generations to the achievement of a simple task by a carefully standardized procedure; and testing the later generations in respect of their facility in mastering the task by comparing them with the earlier generations, with rats of the other half-stock, and with rats of stock newly obtained and not trained to a specific task of any kind. If Lamarckian transmission takes place in any degree, it should be revealed in the later generations (even though it occur in very slight degree only) by some increase in facility of learning; for, in the procedure adopted, slight differences of facility can be measured.

I will describe roughly only one of the two procedures, referring those interested in the technicalities to the published reports.[1] The rat is placed in a large tank half full of water. He can escape only by swimming to either one of two sloping gangways. One of these two gangways is brightly illuminated and so connected that, as soon as the rat touches it, he receives an electric shock. The shock is strong enough (as judged by the rat's behaviour) to provoke some fear. Upon repetition of the procedure, every rat learns, sooner or later, to avoid always the brightly lit gangway and to escape quickly by the other. The question, then, is: Do the rats of the later generations of the trained half-stock learn this avoidance on fewer repetitions of the procedure than are required by other rats?

The training has been given to rats of thirty-seven successive generations; and the general result is that the rats of the later generations of the trained half-stock do learn to avoid the shock more quickly, with fewer repetitions of the shock, than any other rats. The difference is so large as to be unmistakable;[2] and the number of rats dealt with seems to be so large that the difference cannot be attributed to happy (or unhappy) chance. It seems to be good evidence of Lamarckian transmission. I add that the procedure adopted with the other half-stock has given similar, but less clear, evidence of the same general nature.

I do not ask that this experiment (which is still going on to further generations) shall be regarded as giving

[1] *British Journal of Psychology*, April 1927, January 1930, and October 1933.
[2] Roughly, the rat of ancestry untrained to this task requires more than one hundred repetitions of the shock before learning to avoid it; while a rat of the later generations achieves complete avoidance after some fifteen to twenty shocks only.

a final and positive answer to the much-debated question of Lamarckian transmission. I write this article chiefly as a warning against hasty application of this conclusion. On an earlier page I pointed out the far-reaching theoretical bearings of the Lamarckian question. Either answer to the question, the negative or the positive, must have practical bearings that are equally important, and must affect our attitude towards a number of political problems and social programmes. Of these, the most important is the problem of preserving and promoting the qualities of our human stocks.

Of recent years many men of science have warned us that the human race is threatened with degeneration. Against that danger we may be able to take effective measures, if only we can achieve sufficient understanding of the laws of heredity and variation. We may even be able to devise and put into practice measures which will bring about some improvement of its qualities. And such improvement is perhaps essential for the flourishing of our civilization: for it appears only too clearly that, as at present endowed, the human race has hardly sufficient intelligence and moral quality to be entrusted with the enormous forces which science is placing at its disposal.

There are two great parties among those who realize the importance of measures to conserve and improve human qualities. The one party, known as the advocates of euthenics, would rely upon improving the conditions of life for men in general, upon improved public hygiene and, especially, upon more intensive and widespread education. They assume, explicitly or implicitly, the truth of the Lamarckian principle, believing that, if, by means of better conditions of life and better training, we improve the qualities of each generation of men, we shall thereby improve the inborn constitution of the race.

The other party, those known as advocates of eugenics, have for the most part accepted the neo-Darwinian denial of the transmission of modifications. They point to various forms of selection going on among human populations, especially those that live under the conditions of civilized society. They maintain with much force that the total result of these forms of selection must be gravely adverse, must make, not for improvement, but for rapid deterioration of the human stocks subjected to them. They assert that the measures upon which the euthenists rely, far from promoting the qualities of the race, must, in the long run, prove to be powerful agents of deterioration. They point to the fact that modern medicine and hygiene now keep alive millions of weakly children who, in the conditions of former ages, would have succumbed in the early years of life. They point out that our free, but costly, educational processes, our free hospitals and parks and playgrounds, while they benefit the children of the poor, throw a heavy burden of taxes on the more successful classes and thus play no small part in inducing among them that restriction of the birth-rate which is one of the most significant and fateful phenomena of the present age.

Thus acute opposition has arisen between the euthenists, who demand improvement of environment, and the eugenists, who assert that all such improvement must be powerless to prevent racial deterioration unless the present adversely selective processes can somehow be stopped, or counteracted by favourably selective processes.

In this article I am chiefly concerned to protest against any hasty application of my experimental results, or of any similar evidence, as a weapon against the eugenic movement. If the reality of Lamarckian transmission has been, or should shortly be, demonstrated, should that encourage us to throw aside the warnings of the eugenists and to proceed confidently with a purely euthenic social programme? To my mind the answer is clearly in the negative. If Pavlov's allegedly successful experimental results had been able to withstand all criticism and had been abundantly confirmed, I think we should have been justified in giving adhesion to a purely euthenic social programme. For we might then have hoped that the improving influence of improved environment and education would be so great as to overpower and swamp all the influences of adverse selection which characterize our civilization. On the other hand, if Lamarckian trans- mission is real, but as slight and gradual in its effects on the race as is indicated by my experiments, then we must still give heed to the eugenists; we must aim at a social programme which shall conform as fully as possible to the demands both of the euthenists and of the eugenists. For we must note that the training of thirty-seven genera- tions of rats was required to produce a modest increase of racial facility, a slight and highly special racial improvement. Now, thirty-seven generations of mankind mean a period of a thousand years at least. And it may well be that the adverse selective influences, on which the eugenists insist, may be so powerful as to overcome and neutralize the racial benefits which the realization of the most ideal euthenic programme could achieve in the course of ten centuries.

Demonstration of the reality of Lamarckian trans- mission should, then, not blind us to the dangers of a purely euthenic programme that neglects the considerations advanced by the eugenists. It should rather lead to the resolution of the conflict between euthenists and eugenists, enabling us all to unite upon a common social programme in which due weight shall be given to both euthenic and eugenic measures. And it should have this further moral advantage: it should enable us to go forward with confidence in the future of our race; for if the Lamarckian theory is true, man may fairly hope to achieve by his own efforts a higher level of development, on which he may reap in full the immense advantages of those new resources which science is bringing within our reach in ever fuller measure.

A well-founded belief in Lamarckian transmission must have, in yet another way, a great moral influence. What belief could stimulate so powerfully to individual effort at self-improvement—moral, intellectual, aesthetic, and bodily—as the belief that every improvement we achieve in our own personalities will be in some measure, however slight, transmitted to our children and our children's children and by them perpetuated so long as the human race shall endure.

XII
WORLD CHAOS[1]

THE civilization of any people reflects the state of its knowledge, and is in a large measure determined by that knowledge. If, then, we have reason to be profoundly dissatisfied with the state of our civilization, we shall do well to consider whether there is not some radical defect in our knowledge, more especially in the systematically organized part of our knowledge which we call Science.

It would not be true to say that our Western civilization is founded upon science; for its deeper foundations are the curiously blended traditions of Christianity and of the classical world of Greece and Rome. But modern science, the science that dates from the Copernican revolution, the science founded by Copernicus, Galileo, Descartes and Newton has profoundly modified it, has given it a highly peculiar quality, a quality that tends to overshadow, to obscure and even to destroy those ancient foundations. For the most distinctive quality of our present civilization is that it undergoes perpetual and rapid change and that its ideal is progress rather than stability; and this quality is the gift of science.

If all seemed well with us, if peace and material prosperity seemed assured to all nations, the present state of our civilization would nevertheless justify some misgivings; we might well ask: Can changes, so great, so rapid, so continuing, be compatible with stability? We are building up an immense and very complicated superstructure. May it not be top-heavy? May not the foundations be

[1] A lecture given before the University of Manchester on the Ludwig Mond foundation in May 1931.

crushed and crumbled, until the whole structure erected on them must collapse?

It was, no doubt, a sense of this increasing top-heaviness of our civilization that prompted an English bishop recently to suggest that science should take a holiday, that for half a century, at least, all scientific research should be suspended while we consolidate our gains and make sure of our foundations. The suggestion is an impracticable one. We cannot stop and stand still even if we would; we are committed to further progress or, at least, to further change. The knowledge and power which science has given us must inevitably carry us on with a frightful momentum. We are like men in an aeroplane crossing a great ocean; and the driving power is science. To cut off the engine in mid-ocean could result only in disaster. Though we know not where we are, nor whither we are going, we must keep on, hoping for a happy issue, applying the discoveries of science to the best of our ability and our wisdom.

A complete cessation of all scientific research would, no doubt, result in a slower rate of change than is likely under the influence of continued and increasing research. But the process of change has a momentum of its own which must carry it on, perhaps at an accelerating pace, even without the influence of further research. The further diffusion of the knowledge we already have, the wider technical application of that knowledge, the many social and political changes already in progress are consequences of the application of scientific knowledge and of the undermining of old traditions and beliefs by that knowledge;—all these consequences will inevitably continue to unfold themselves, producing new complications, new problems, new dangers, as well as new benefits. Consider the knowledge of methods of birth-control. This knowledge, now let loose upon the world irrevocably, may within a few generations result in changes of the composition of the world's populations far more violent and momentous than any recorded by history, than any resulting from the most devastating plagues, the most destructive wars, floods or other catastrophes. And the changes of population that may well result from this new knowledge are likely to be far more profound and lasting, and only too likely to be far more disastrous, than any previous changes of population; because, while in the main those former changes were non-selective, affecting all parts and degrees of each population equally, or, if selective, then, in the main, favourably selective, the changes due to the new factor are likely to be highly selective and so far as can be foreseen, in the absence of public regulation of some sort, selective in a highly unfavourable sense.

Consider the influence of the means of easy transportation of human beings in large numbers over great distances. This factor has been in operation for only a short period; yet already it has effected changes in the distribution of races and peoples in comparison with which all earlier migrations vanish into insignificance.

Consider the effects of 'radio' and of the 'cinema', now just beginning to make themselves felt in all parts of the world. Who can foresee the extent and nature of their influence? All we can certainly foresee is that the effects must be very great and world-wide.

Consider the effects of the spread of large-scale machine-production among the populations of Russia, India and China. Already the inception of the process in Russia is producing profound uneasiness in the nations already industrialized; just because the world-effects must be so vast and are so unpredictable.

The proposals of Mr. Gandhi and of other enthusiasts for the simple life come too late. It is impossible to forswear the methods of machine civilization and to return to the simple economy of hand-production and of agriculture with primitive tools, if only because, by the aid of machines driven by steam and electric power, we have multiplied immensely the numbers of mankind. Mr.

Ghandi's ideals, if they could be put into practice throughout India, would entail the reduction of the population by some two hundred millions. The lives of vast numbers of mankind are increasingly dependent on the unremitting diligence of the highly trained experts who maintain in working order, by the aid of scientific knowledge, the vast systems of power-supply, of transport and of communication, and only in less degree upon those who maintain sanitary services, manufactures, and the protection of animal and plant life against pests and plagues which under modern conditions would, if uncontrolled, threaten the destruction of our principal food-supplies. The spread of infectious diseases of animals and plants, as well as of human beings, from one continent to another has been one of the most troublesome consequences of the modern facilities of transportation.

Our civilization is top-heavy: this is a defect and a danger which we can do little or nothing to remedy. The foundations were laid long ago, the superstructure exists as a going concern and we must keep it going. We cannot refuse to use the immense resources which science has put at our disposal.

A second characteristic of our civilization is no less a defect and a danger than its top-heaviness; namely, it is lop-sided. Now a top-heavy structure may stand firmly erect if it is symmetrical and well balanced; but, if it is both top-heavy and lop-sided, disaster threatens. And not only is our civilization both top-heavy and lop-sided; it grows more top-heavy and lop-sided with every year that passes; consequently its stability becomes more doubtful, and the danger grows more threatening. I have said that we can do little or nothing to make our structure less top-heavy; fortunately it is possible to make it less lop-sided, and thus to give it greater stability.

This second defect and the remedy for it are the topics of this brief discussion.

The top-heaviness of our civilization is due to the rapid

development of science; its lop-sidedness is due to the lop-sidedness of our science. Our civilization reflects the state of our knowledge; and especially it reflects it faithfully in respect of the lop-sided state of our science.

That our science is very lop-sided is indisputable, a matter of common agreement. Since the time of Galileo, physical science (by which I mean the sciences of the inorganic or physical realm) has advanced at a constantly accelerating pace. The sciences of life have lagged far behind. Until the middle of the nineteenth century they were rudimentary, concerned merely with description and classification. The work of Charles Darwin gave them a fillip. For a time it seemed as though they were about to progress rapidly; it was even confidently prophesied that, as the nineteenth century had been the century of physical science, so the twentieth century was to be the century of the biological sciences, was destined to go down in history as the great age of biological and psychological discovery. We are now near the end of the first third of the century, and it seems very unlikely that this prophecy will be realized. Physical science still accelerates its progress. The biological sciences limp painfully behind. If this is true of biology proper, still more is it true of the sciences of man and of society. We talk of psychology, of economics and of political science, of jurisprudence, of sociology and of many other supposed sciences; but the simple truth is that all these fine names simply mark great gaps in our knowledge, or rather fields of possible sciences that as yet have hardly begun to take shape and being. The names stand for aspirations rather than achievements; they define a programme, they vaguely indicate regions of a vast wilderness hardly yet explored, and certainly not mapped, regions in which chaos still reigns, yet *regions which must be reduced to order if our civilization is to endure.*

The grounds of this lop-sided state of our science it is easy to indicate. First, the physical sciences are the easier: as M. Bergson has so forcibly pointed out, their problems are such as our minds are adapted to deal with. The mental capacities of the human species have evolved in the course of a struggle for existence, in which struggle the first condition of survival has been effective dealing with the material environment; and in spite of all the fine things said about the scientist's pure love of truth, the truth seems to be that the main spur to the development of physical science was man's practical need of understanding and control of his physical environment.

The success of the physical sciences in bringing such understanding and control has been and still is the second great ground of the predominance of those sciences in our field of knowledge. Physical science has brought such great and obvious benefits to mankind that it commands the respect, the admiration and the ungrudging support of all the civilized portion of the human race. I need not dwell upon these benefits, the multitude of conveniences, comforts and luxuries. Look about you, consider the contents of your breakfast table, or your parlour with its electric light and radio-outfit; or penetrate to remote jungles of Malaysia and see the naked savage using lucifer matches from Japan, cotton cloth from Manchester, pottery from Staffordshire, a rifle from Hartford, Conn., and, perhaps, spectacles from Birmingham.

The prestige accruing to physical science from the practical benefits it has conferred is the greater because those benefits are in the main obvious and unquestionable; and the fact that they have resulted directly from scientific research is easily grasped by the common man. The steam-engine, the motor-car, the rifle, the radio, the telegraph and telephone, and a thousand other mechanisms in common use perpetually remind the common man of his debt to physical science. But, when he eats cheap and good and varied food gathered from remote parts of the world, nothing reminds him that biological research has contributed greatly to make this possible. Still less does he realize that, if his expectation of life and health is

prolonged by some twenty years beyond that of his grand-
father, this is the consequence of biological and medical
research. And, when he reaps the benefits of living in a
well-ordered peaceful commnity in which his rights are
protected against the rapacity of rulers and the greed and
lust of his fellows, in which vastly complicated systems
of finance, of education and of parliamentary representation are maintained in smoothly working order, he is a
very exceptional person if he at all realizes that all this
has been rendered possiblê only by a vast amount of
thought and discussion and research in the field of the
sciences of man and of society.

A third ground of the backwardness of the biological
and human sciences has been the opposition of the Churches.
The Church quickly learnt to adapt its doctrines to the
Copernican revolution and to the teaching of Newton;
and it welcomes with open arms and loud acclaim the
Einsteinian revolution.[1] But the Church, though it has permitted the study of some of man's cultural achievements,
such as language and literature, has frowned upon the study
of his beliefs, his superstitions and his religions, his magic
and ancient customs; and especially it has opposed all more
direct approach to the problems of human nature, as well
as all biological studies (such as the dissection of the
human body and the comparative study of men and
animals) which tend to reveal man as a part of nature.
And, let me remind you, our universities have been,
and still are, largely controlled and shaped by the Churches.

A fourth ground of the rudimentary state of the biological
and human sciences is perhaps the most important at the
present time.

The long start and the relatively advanced state of the
physical sciences, together with the great prestige accruing
to them from their brilliant successes, have worked to the
great detriment of the sciences of life.

[1] I see the Einstein principles are invoked just now to render
more easily credible the story of the Ascension.

And this in two ways. First, the methods and principles that have proved so successful in the physical sciences have been accepted by a great majority of men of science as alone valid for all sciences; and the account of the world rendered by physics has been accepted by most men of science and by many philosophers, from Spinoza onward, as literally true. The work of modern philosophers has consisted very largely in fruitless efforts to reconcile with the mechanistic principles of physical science, and with the account of the world rendered by it, some belief in human values and in human efforts to conserve and augment these values. The biologists, accepting physical science as the model of all true science, have for the most part been resolutely blind to the abundant evidences of the causal efficacy of mental activity in the organic world. This general acceptance of the mechanistic account of the world has hampered and perverted all biological and, especially, all psychological research. And, like a dark cloud, it has enveloped and overwhelmed the popular mind; has impressed and oppressed it.

The depth of this impression and oppression has been vividly illustrated by the world-wide applause which has greeted recent announcements by physicists that physical science has been in error. When physicists, like Jeans and Eddington, have recanted the major errors of the scientific dogma, the world at large, instead of turning and rending them with reproaches, instead of crying 'Then why have you physicists so long oppressed us with your nightmare dogmas?' still grovels before physical science and regards these recantations of error as announcements of great scientific discoveries. So great is the prestige of physical science in the popular mind!

A vicious circle has been established. Physical science, having obtained a long lead and a vast prestige, continues to increase her lead and her prestige at the cost of the sciences of life; for, as we know, nothing succeeds like success. The universities provide palatial accommodation

13

and unlimited equipment and a multitude of academic posts for the workers in the physical sciences. In addition, great industrial corporations spend vast sums on physical research. It is officially estimated that in the United States alone research (almost exclusively in the physical sciences) is subsidized by industry to the extent of five hundred million dollars a year, a sum considerably greater, I suppose, than the combined incomes of all British universities. Lord Rutherford (*The Times*, 9th June 1931) is quoted as saying that tens of thousands of men are to-day engaged in research directed to improvement of the motor-car; and it has recently been alleged that, while in America two hundred chemists are devoting their energies to the discovery of more deadly poison gases, this country is spending £200,000 a year in the same noble cause.

Consider now the state of our civilization. It reflects, as I said, the state of our science; it is therefore characterized by its multitudinous applications of the discoveries and inventions made by, or made possible by, the physical sciences, applications which have transformed our world and our relations to one another, political, social and economic, national and international. The total effect is commonly described, not quite accurately, yet not quite unjustly, as an increasing mechanization of all our civilization.

The principal feature of the change wrought by physical science is that we have at our disposal vast stores of physical energy; which energy we have applied in two ways. First, as a substitute for the muscular energy of men and domestic animals in the production of the fundamental necessities of human life, food, shelter and clothing; *with the immediate consequence that the population of the world has multiplied as never before*. And it is important to notice that this very rapid multiplication has taken place not only in those peoples who have developed the physical sciences and their applications, but also, in only less degree, among the vastly more numerous peoples who have had no active share in that development. I remind you, merely as an illustrative instance, of the fact that the population of India was multiplied roughly threefold during the nineteenth century.[1]

Secondly, we have produced a multitude of comforts and conveniences which have become so intimately woven into the texture of our civilization that they are now essentials; the deprivation of them would cause not only much discomfort and suffering but also the breakdown of the whole complex structure of our civilization; without them we should starve and die by millions in all parts of the world. That is to say, the discoveries made by physical science have greatly increased the numbers of mankind and have added immensely to the complexity, the delicacy, the intimacy, the vital importance of the relations between men and between groups of all kinds.

In all former ages the relations of man to man and of group to group (the civilized no less than the savage) were governed by custom and tradition, law being merely the formal recognition of custom and tradition. The family, the clan, the tribe, the kingdom, the feudal system, the parliament, all such institutions expressed and were adequately governed by old traditional loyalties. Under the vast complexity of modern conditions, this old traditional wisdom is utterly inadequate to regulate our relations. We are compelled to try to live by the light of science; and alas! we have no science to guide us. *The physical science which has produced this new complexity can give us no guidance whatsoever in our difficult task of coping with it.*

Consider our plight. The family, the most deeply rooted of all our traditional institutions, the foundation of all the rest, decays; and we are threatened with general deterioration of the peoples that have created our Western

[1] I have discussed these facts and their bearings at some length in my *Ethics and Some Modern World Problems*, London, 1924.

civilization, if not their actual extinction or substitution by other races.

Here, as in so many other matters, America leads the way and points the path along which the rest of our Western civilization seems destined to wander.

Our most approved political institutions are much blown upon; representative democracy based on universal suffrage, the ideal for which the pioneers of the nineteenth century enthusiastically strove, is no sooner realized than it proves so disappointing and inefficient that we see great areas reverting to various forms of tyranny.

Here America cannot claim to be in the van of progress! The palm goes to Italy.

Our so-called international law has been proved by the Great War and found to be an ineffective sham; and the League of Nations vainly strives to bring about disarmament on the very questionable assumption that general disarmament, if it were possible (which it is not), would prevent war. Meanwhile we spend vast sums in preparation for war, sacrificing in the process the lives of our finest young men; and we drift towards new wars that threaten to make an end of us and all our works.

The Churches keep crying aloud their old stories and their old exhortations, but the people heed them less and less. In education we are all at odds; the only ideal that seems to make an effective appeal is that of keeping the children in school a year or two longer in order to make more jobs for their elders.

In the economic sphere the tragic absurdity of our predicament reaches its climax. Through the aid of physical science our powers of production have reached a very high point of efficiency; an efficiency such that, if the whole machinery of production could be set working at full speed, every human being might be lapped in luxury of the most elaborate kind, at the cost of a modest expenditure of human energy. Yet the whole world is poverty-stricken in various degrees, and even in America there are said to be at least six (perhaps eight or more) million workers out of work[1] and a considerably larger number suffering serious deprivations.

If it be said that this is the nemesis of capital so confidently foretold by our Socialist friends, we cannot forget that in various countries the Socialists have attained to political power and have shown themselves unable to effect any remedy. While in Russia the Soviets are making a vast experiment in communism which, however successful it may prove in the purely economic sphere, seems likely to be put through only at a frightful cost of suffering and servitude, a cost that may well prove excessive and disastrous.

We live, then, in an age of grave social disorder and threatening chaos; and it is in the main due to science. What then is the remedy? The remedy for the ills of science is more science, more knowledge systematically organized. But what sort of science? Physical science has been the main agent in producing our chaos; and physical science can bring no remedy. Suppose that physical science should continue its brilliant and accelerating course; suppose that it should discover heavenly bodies a million times more remote from us than any yet observed; that it should enable us to see and hear at any moment what is going on at any point of the earth's surface and to travel thither in a few seconds; suppose that it should put at each man's service (on the average and in principle) energy equal to that of ten thousand slaves, instead of only fifty as at present;[2] suppose it should invent explosives and gases a million times more destructive than any now available; suppose that another Einstein should convince us all that space is zigzag or that time is square. Should we be any happier or safer for any or all of these advances? Consider what would happen if some brilliant physical

---

[1] To-day (1934) the number is said to be anything from twelve to fifteen millions.

[2] It is estimated that in America this number is about 150.

discovery were to put us in free communication with inhabitants of Mars not very unlike ourselves. We should forthwith be absorbed in efforts to prepare some ray with which to blast them from their planet, in the 'purely defensive' warfare which our fearful imaginations would anticipate. And if some physicist were to realize the brightest dream of his kind and teach us to unlock the energy within the atom, the whole race of man would live under the threat of sudden destruction, through the malevolence of some cynic, the inadvertence of some optimist, or the benevolence of some pessimist. *I submit that no such discoveries, nor any others that physical science could possibly make, could avail to remedy our condition.* I would go further and assert that every step of progress physical science may make in the near future can only add to our dangers and perplexities: for every step of such progress must increase the top-heaviness and the lop-sidedness which are the radical faults of our civilization.

I have no wish to belittle the achievements of physical science. They are immense and altogether admirable. I am concerned only to bring home to the minds of my readers the indisputable fact that the very successes of physical science, leading as they have done and inevitably must do, to rapid and violent changes in all our modes of living, are producing a state of affairs which is ever more unstable and dangerous, which urgently needs some large-scale corrective such as physical science, no matter how successful, is, in the nature of things, unable to supply.

In this appalling situation, in face of this dread prospect, we continue actively to augment the sources of our trouble; we continue deliberately to increase this lop-sidedness of our civilization, we devote more and more of our resources to physical research. 'Those whom the gods would destroy they first make mad.'

It is only the biological and especially the social sciences founded on biology that can save us.

My thesis is that, in order to restore the balance of our

civilization, in order to adjust our social, economic, and political life to the violent changes which physical science has directly and indirectly produced, we need to have far more knowledge (systematically ordered or scientific knowledge) of human nature and of the life of society than we yet have. First we need to know the truth about differences of fundamental constitution between races and individuals. Are these very slight and unimportant, as one large school of opinion confidently asserts? Or are they, though difficult to define, of profound importance and very difficult to modify, as others of us believe, though at present we cannot adduce conclusive evidence? Is it true that some existing human stocks are far more capable than others of producing and maintaining a high civilization? Can we hope that, with or without socially directed effort, the existing races of man are likely to advance in respect of the qualities that make for a high level of civilization? Or is a general decay or falling off in quality probable or inevitable? What measures can be taken to promote the one possibility and make the other more remote? Are we at present covering over, by means of improved hygiene, education, and general conditions of living, a subtle degenerative process affecting perhaps all the more civilized part of mankind? May improved conditions of life, with improved training and efforts at self-improvement, effect improvement of the race, or secure it against deterioration, as the Lamarckian theory, if it could be substantiated, would lead us to believe? What are the effects of the cross-breeding of the various human stocks? Are they all good or all bad? Or are some good and some bad; and if so which? All these are questions, profoundly important for the future of mankind, to which biology at present can give no sure answers. Briefly, then, we urgently need a well-founded theory of evolution such as at present we lack; and we need knowledge of its detailed application to the human race.

Secondly, we need the development of the social sciences,

economics, politics, jurisprudence, criminology, penology, history, social anthropology, and all the rest, for our guidance in all social and political problems, in face of all of which we stumble blindly along amidst a chaos of conflicting opinions. And all of these need for their foundation some sure knowledge of the constitution of human nature and of the principles of its development; in other words, a sound psychology. Take the great question of perennial dispute—the most desirable politico-economic organization of peoples; should it be democratic or authoritarian? Should it be individualistic, socialistic, or communistic? The answers to all such questions depend upon the assumptions we make (and, in the absence of sure knowledge, are free to make) about human nature. Will the citizens do their duty and lead the strenuous and co-operative life without the spur of competition and the threat of economic misery? No one knows, and we all make (for the most part implicitly only) what assumptions we please, and shape our answers and our political applications accordingly.

Consider the sphere of international relations. National-ism, it is generally agreed, has been the greatest moulding force in the history of the nineteenth century; and it has become acutely accentuated in recent years. It is widely denounced as the greatest evil of the present time. Yet others, like the late Duke of Northumberland, deplore the decay of nationality and of patriotism; and where, as in India and China, nationality is as yet merely an ardent aspiration, the very persons who most loudly denounce nationalism in Europe as a child of the devil exhort us to encourage and sympathize with the nascent but pro-foundly disturbing nationalism of these ancient civiliza-tions. And, again, when the late President Wilson pro-claimed the great principle of self-determination for all peoples and succeeded in applying it so far as to Balkanize half of Europe, his supporters and accomplices were in the main the same persons who now so loudly denounce nationalism. What, then, is this greatest of all modern forces, nationalism? And what are patriotism and nationality, and what their relations to nationalism? These are questions that must receive clear answers before we can hope to emerge from this welter of confusions and contradictions. And all these are psychological questions? Whatever else this much-discussed force, nationalism, may be, it is in some sense composed of a multitude of energies that reside and operate in the breasts of human individuals; and its workings cannot be understood until we have at our command some well-founded psychology, both individual and collective. Again, economics is, or should be, a science that deals with the desires of men, their strivings and their intellectual operations, the processes by means of which they strive to attain satisfaction of those desires—primarily and fundamentally a psychological science. Yet one half its official exponents deny this simple truth, while the other half pays it lip-service only. But of this more later.

Or consider jurisprudence and the allied problems of criminology and penology. Is it not true that all these complex blends of science and philosophy are concerned to regulate, to direct, to order and reform the conduct, the thoughts, the intentions, the feelings and the actions of men and societies? And how shall they do this unless they truly conceive, however inadequately in detail and in special cases, the general principles of human motivation? It is pathetic to read the works of a great jurist like the late Sir James Stephens, to see this powerful intellect struggling vainly with its complex problems; vainly, because he starts out with a fundamental assumption about human motivation, the assumption of hedonism, widely current in his time, yet radically false and misleading. It is like watching a lion struggling in the toils of a strong net.

I have said that at present we have no such sciences. The statement may seem extreme. It is true we have some beginnings of biology; notably a vast mass of 'data', some highly specialized departmental studies, and some successes in the sphere of empirical medicine. But we have no biology that can serve as the basis of the social sciences we so urgently need. And the social sciences themselves are merely a faulty sketch of a programme.

I cannot pretend to examine each of them in turn. Let us glance very briefly at that one which has the longest history and the most considerable body of students, and which has the most immediate bearings on social practice. I mean economics. Will any one affirm that economics is a science; that it is anything more than a frightful mess of statistics and highly questionable theories? The supreme test of a science is its power of prediction. It would be too much to say that the predictions of the economists are always wrong. But, if some of them have not been wrong, is it not merely because, among a multitude of predictions, some must hit the mark according to the law of chance?

Apply the less exacting test of power to explain after the event, and what do we find? A sheer chaos of opinions, the highest authorities directly opposed to one another and a multitude of smaller fry with their own versions. Take the question of the grounds of the present world-wide economic slump, or almost any other economic phenomenon you please; and it is always the same. Could there be a greater diversity of opinions and acuter opposition of authorities? Take the single question, How far is the distribution of gold a principal factor? And when we come to recommendations for economic betterment we find chaos raised to the nth power.

Consider the chaos of opinions and recommendations concerning the gold-basis. The necessity of a gold-basis has been for generations an accepted dogma. Only a few of us have wondered whether it was not merely a superstition. And now at last it is blown upon from many quarters. Economists of the highest repute tell us that it is not only a superstition but a most pernicious one, the main ground of all the economic sickness of the world in general and of Great Britain especially.

In every branch it is the same story. The books on economics are full of iron laws, inexorable laws. Yet in any true sense is there an economic law anywhere in sight? The very language the writers use is hopelessly loose and confused. Let me cite one instance which points to the root of the trouble. Arnold Toynbee, comparing economics with physics, wrote: ' But the economist has to deal with facts which are far more complicated, which are obscured by human passions and interests, and, what is still more to the point, which are perpetually in motion.' What does he mean by saying that the physical facts are at rest and the economic facts in perpetual motion? We can only guess. But more significant is the statement that the economic facts are obscured by human passions and interests. He should have said rather that the essential economic facts are 'human passions and interests'. The economist constantly speaks as though there were a realm of facts and laws which he might reduce to order, if only there were no human passions and interests to pervert his facts and laws and to frustrate his praiseworthy efforts to reveal the facts and formulate the laws. Toynbee is out of date, you will say. But when I turn to the most authoritative contemporaries I find the same state of affairs. Mr. A. Loveday, the economic adviser to the League of Nations, has recently published a book. In that book he frequently writes of economic 'forces and tendencies' that are said to operate with greater or less force in this or that country. It seems clear that when statistics reveal any continuing change, he postulates a 'force and tendency' as the cause of that change. But what the force and tendency may be he does not stop to inquire. He, like Toynbee, merely querulously implies that human passions and interests are unwelcome disturbing factors which made difficult the economist's task of defining these 'forces and tendencies'. He writes: 'I tried to lay bare certain forces and tendencies which owe their existence mainly not to conscious and concerted human effort, but to unco-ordinated changes in individual action and to the natural growth of wealth.' He speaks of various forces and tendencies being checked and arrested, as though they still continued in being, while counteracted by others, in analogous fashion to physical forces.

I turn to Sir Josiah Stamp, and I am happy to find that he insists, as I am doing, on our urgent need for economic science. 'We are in real peril, and a serious breakdown of our economic society is far from being impossible.' 'To-day is the day when of all times the drive is wanted in economics . . . we cannot have too many hard and brilliant thinkers in that field, too much money poured into research, and too much patient self-denying effort to advance the science.' And he rightly points out that mere accumulation of statistical facts is of no avail. 'We have vast masses of facts . . . in the coal industry in England every fact has been meticulously known . . . no industry . . . is more fully documented than the coal industry, but unfortunately the psychological conditions behind it . . . still have their influence on it.' Here you see again the same assumption of a realm of economic facts and wickedly disturbing 'human passions and interests' frustrating the good intentions of the economist to make a science; the assumption of laws that *might be* iron and inexorable if only human passions and interests would not interfere. And the same old fundamental error of the economists comes out, when in other essays the same authority repeatedly distinguishes between, on the one hand, economic facts, and, on the other hand, our thinking about them and our personal attitudes towards them. There by implication he denies that our thinking and personal attitudes are economic facts; whereas they are in truth the most important of all economic facts. Economics deals with values, and there is nothing good or bad (nothing of value), but thinking makes it so. How absurd, then, this pretence that economics can abstract from the human factors and discover a realm of facts and laws that may be independent of this cruelly disturbing factor, human nature and its vagaries!

The assumption of an economic robot dates from the early days of the classical political economy; it still survives in the implicit assumption that the laws of economics would be valid if only men were such robots.

It is the basic error of most economists; the assumption of an iron man governed by iron laws; a robot so simple that it may be left out of account after we have made a few deductions about its working. It is no wonder that economics, with its iron laws true only of iron men, is often called the most inhuman of the sciences; for economics strives constantly to ignore and abstract from the human factors which are the most important of the facts and which express the most important of the laws of which it must take account. Economics is surpassed in its inhumanity by its special branch, finance or banking. Do I need to illustrate the fact? Perhaps it may suffice to refer again to the banker's superstition of the gold-basis. Or take a comment made by one of our great dailies[1] on an international banker who has recently ceased from troubling, one who had played a large part in bringing about the present state of the world. It wrote that he knew everything about money but nothing about the imponderables. By that it clearly meant that he knew nothing about the human factors; which was to say that all his opinions were worthless, probably, in many cases, far worse than worthless. For what is money apart from the human factors? It is literally nothing. The value of money, like all our other values, is a function of human nature. What a shocking state of affairs! That the welfare of many millions of men throughout the world should be at the mercy of a man who knew all about money,

[1] *Morning Post.*

and nothing about the imponderables. No wonder the world is in chaos!

If we look round to pick out the really notable contributors to this field, those who have done something to illuminate it, we find it is those who do not neglect the imponderables, who treat them, not as annoying and undesirable complications of some alleged purely economic facts, but as the most essential, the all-important, economic facts. Such was Walter Bagehot. Such in our own day is Mr. J. M. Keynes. I look at a recent short article from his pen discussing tariffs, and I find thirty references to the imponderables. And one of these imponderables is mentioned seven times, and would appear to be, in Mr. Keynes's view, the all-important factor in the contemporary economic world, and the lack of it the prime factor in maintaining the distressful position of Great Britain, namely, *confidence*.

Here we have a pointer; it points directly to the conclusion I am urging. What is this imponderable, so overwhelmingly important in the economic sphere—this *confidence*? Surely it must be worth while to study it; to try to learn something of its nature and conditions and laws. For surely it is a natural phenomenon that has its conditions and conforms to laws. And the same may be urged of all the other imponderables mentioned by Mr. Keynes and of many others which are no less important for economics and for all the other social sciences.

But does Mr. Keynes or any other economist or political scientist propose to inquire into the nature and conditions of confidence? Not at all! Confidence remains an utterly vague word. It is clear that it stands for some natural phenomena of immense importance; it is clear that the conditions under which confidence ebbs and flows are vastly complex. It is clear that the phenomena and their conditions are deserving of the most intense study, that, in fact, all economists and statesmen should be chiefly concerned to learn all that can be learnt concerning them.

But, no; while we spend millions on the search for new and deadly gases, and have tens of thousands of experts engaged in research for the improvement of motor-cars, we spend not a penny on research into *confidence*, which is more vitally important for the very life of our civilization than all the multitudes of physical researches now going on in every civilized land; more important than all that these researches can possibly reveal to us in the course of a century.

I submit that economics is not a science, but, rather, a chaos; and that, in spite of the great amount of work done in the field, it must remain a chaos until it can found itself on some systematic knowledge of these imponderables. And the same is true of all the so-called social sciences. As Mr. Loveday truly says: 'It is well to remember when conditions are bad, men are without work, and children short of food, that the malady from which society is suffering is plain ignorance.' But not ignorance of facts of the statistical kind, of which, as Sir J. Stamp tells us, we have multitudes, a plethora. We lack rather such knowledge of the imponderables as will enable us to interpret these facts.

It is the same story with all the so-called social sciences. They are backward, rudimentary, not only because they are starved and neglected, but also and chiefly because they have not for their foundations any science of the imponderables, those human energies with which they are almost exclusively concerned; and, because the exponents of these sciences, so far as a large proportion of them are concerned, are not even aware of this need, this lack of an indispensable foundation.

Consider for one moment so-called political science, or the science of government. Is it not almost as chaotic as economics, as incapable of prediction, of explanation and of trustworthy precepts? And is not the ground of its defects the same? When it aspires to predict, it has nothing more than the bald statement that history repeats itself—or does not repeat itself—according to the fancy of the author. When it seeks to explain, it invents *ad hoc* various strange instincts. When it is a question of prescribing, it can offer only a few vague and abstract generalities.

Here, then, is the only road to remedy the parlous and ever more dangerous state of our civilization. We must actively develop our social sciences into real sciences; and, in order to do that, we must first create a science of the imponderables, in short, of human nature and its activities. But perhaps that is impossible. The school of thought predominant in this country asserts the impossibility. They tell us, by implication, if not in so many words, that all we can ever hope to learn of the imponderables is to be found in the works of Plato and Aristotle. This seems to me a doctrine of despair. For we have tried the intensive study of these great authors for many generations; and it has not met our needs. Such study is admirable as a method of attaching our civilization to its foundations; but of little avail to correct its lop-sidedness, to furnish a basis for the social sciences we need. To despair of achieving systematic knowledge of the imponderables before we have made a serious and sustained effort would be to manifest a craven spirit. What, then, in practical terms, is the remedy. I can give my answer most concisely by suggesting what I would do if I were dictator. I would, perhaps, permit the continuance of physical research by the industrial corporations. But I would by every means seek to divert all our most powerful intellects from the physical sciences into research in the biological, the human and the social sciences; and our universities should be the main seats of such research.

Especially, I would at first concentrate the attack upon anthropology, the science of man, conceived in the broadest way. It would take some twenty years to train the personnel and get them working on their problems. Then, as they began to bring in results, part of the available intellectual energy would be concentrated in the effort to build the social sciences, especially a science of economics, on the basis provided by the anthropological research. On a hopeful view, another twenty years would elapse before substantial progress along this line might be expected. That would bring us to the year 1970, or thereabouts.

Can we afford to wait so long? Can our civilization survive in the meantime? I don't feel sure; but I hope it may: for I can see no alternative measures that offer hope of its salvation.

> The Bird of Time has but a little way
> To flutter—and the Bird is on the Wing.

XIII
OUR NEGLECT OF PSYCHOLOGY[1]

PSYCHOLOGY is at once the oldest and the youngest of the sciences. It is the oldest in that a systematic study of the mental functions of men and animals was made by Aristotle, and with such success that the psychology fashioned by him was more deserving of the name of a science than any other body of natural knowledge at that time. It is the youngest of the sciences in that it is still struggling for recognition in the academic world, and is not yet accorded its due place and influence within our Western civilization. Progress towards such recognition, place, and influence has been made in very different degrees among the various nations; and in this respect Great Britain lags far behind many other countries. The fact hardly needs illustration; I will merely remark that in the universities of Germany there are many professors of psychology and in those of the United States there are some hundreds; while in the whole of Great Britain, there are, I believe, two only.

It is of more interest to point to the consequences and causes of this neglect. I insist first upon what seems to me the most immediately disastrous but least recognized of such consequences, namely in the sphere of economics and industry. The recent strikes, the vast number of unemployed workmen, the persisting stagnation of many industries in Great Britain, contrast painfully with the steady economic recovery of the European nations and the abounding, the almost excessive, prosperity of the United States. It is especially the industrial progress of America that challenges comparison, that urgently raises

[1] Reprinted from *The Edinburgh Review*, April 1927.

the question, Why this difference? Why so much of progress and prosperity there? Why so little here?

The abounding prosperity of America arises from many contributing conditions; but there is one which stands out prominently and is as conspicuously lacking in Great Britain, namely, the harmonious co-operation between management and employees. The fact is becoming known to English observers. Sir Alfred Mond, on returning from a recent visit to America, is cited by the Press as saying: 'Another great factor in American business life is that employers and workmen understand each other to their mutual benefit.' It is not long since America also was the scene of many bitter industrial conflicts; but the art of management has been making rapid progress of late years. The superficial explanation of this American phenomenon is that an industrial prosperity leads to the payment of high wages which produce contentment among the wage-earners. But this inverts the true causal sequence. It is rather the harmony between employers and employed, and the consequent free and effective output of energy by both, which render possible the payment of high wages and large dividends. Industrial experience during the war showed clearly the all-predominating influence of this condition. Relatively small numbers of relatively unskilled, untrained workers proved themselves capable of a vast productivity; because they worked in a spirit of harmonious co-operation with the management, impelled by the same strong motives.

But the further question arises: How has American industry achieved this harmony and this consequently increased productiveness? And the answer can only be that American management has realized, has come to recognize and believe in, the importance of the psychological factors; has learnt to give them careful consideration. The fact might be illustrated in many ways. Many firms employ psychologically trained experts to study their personnel, to sort out the square pegs and put them in square holes, to detect and to remove causes of friction and discontent—those thousand and one small causes which are as so much sand thrown among the wheels of a delicate machine. The universities are recognizing the importance of this work. Harvard, for example, has recently appointed to a chair in her great business school Mr. Elton Mayo, a British subject who has made his name as a distinguished expert in the field of industrial psychology and who, having sought in vain for scope for his special knowledge within the British Empire, has transferred his services to America. Perhaps the most striking illustration is afforded by the work of my friend, Mr. Whiting Williams. Mr. Williams has spent years in making intimate studies of the psychology of artisans and labourers, working beside them just as one of themselves; and now he is invited by Harvard and other leading universities to lecture to their students, to give them the benefit of his observations and reflections. How delicate, how important, how productive such work may be is well shown in Mr. Williams' last book, *The Mainsprings of Men*. In this book, without delaying to explode that monstrous fiction which so long has dominated British economics, the economic man, Mr. Williams reveals the importance of a multitude of obscure and delicate motives working in the bosoms of the horny-handed sons of toil, motives entirely unrecognized by British economists, yet all-important factors of industry.

If the economic superiority of the United States is largely due to the fact that American management has learnt to give due consideration to the all-important psychological factors of industry, how are we to account for the fact? There can be, I submit, only one explanation, namely that in America psychology is everywhere recognized and believed in. It permeates the social atmosphere; it is studied every year by many thousands of young men and women, both in the universities and colleges and in many courses of popular lectures and in popular books and magazines, which, though they may not always impart strictly scientific knowledge, do at least diffuse throughout the people a sense of the importance of the imponderable factors of human life and intercourse. The British attitude towards these imponderables remains one of ignorant scepticism. It was well illustrated by the sneers of the British press when recently M. Poincaré proposed to arrest the fall of the franc and France's headlong course towards national bankruptcy by taking due account of the psychological factors in the situation, in a word, by restoring confidence in the power of the Government to deal with the difficulties which confronted it.

Not only in the sphere of industrial relations, but also in the whole vast realm of economics and politics, the need for the psychological view-point is urgent. British economics, content with its fiction of the economic man, has assumed that, if only men grow rich in material possessions, all else will be added to them in due time—ignoring the thousand and one subtle needs and desires without satisfaction of which the average sensual man, no matter how rich, cannot be happy. Strongly influencing the legislative policies of Great Britain, this economic doctrine has played no small part in reducing our country to its present condition, a condition in which a lamentably large proportion of its population is herded in cities quite unfit for human habitation; is dependent for the satisfaction of its most fundamental and strictly economic needs upon world-conditions over which it has no control; is denied all opportunity for the due exercise and satisfaction of less obvious but no less intrinsic tendencies, the possession of which differentiates its members so widely from 'the economic man'.

All politics and all economics, like all education and most of medicine, must deal at every point with problems that are mainly psychological. It will be long before this elementary truth may gain general recognition in Great

Britain. Yet there are signs of progress. For aught I know there are still British economists who continue to repeat the old nonsense implied by the phrase, 'the iron laws of political economy'. One British economist, however, Mr. J. A. Hobson, has recently discovered the truth and announced it in his last book, *Free Thought in the Social Sciences*, boldly proclaiming that the adoption of the modern psychological view-point by the social sciences is already bringing about a radical transformation of them all. There are other signs of progress in the same direction. The National Institute of Industrial Psychology, which owes its existence to the initiative of Dr. C. S. Myers, seems to have found its feet and to be enlisting the interest of a number of British industrialists. Many books on psychology find British readers. At the last meeting of the British Association, the recently instituted section of psychology attracted, I believe, much larger audiences than any other. The British Psychological Society, which was originated a score of years ago as a small group of persons centred round my laboratory at University College, London, has become a flourishing society boasting nearly a thousand members. In short, a considerable part of the educated public is keenly interested in psychology. Why, then, do our universities continue to neglect so grossly this, the most difficult and the most important of the sciences, and therefore the one that most needs for its development and application that guidance which only the universities can give?

The answer to this question is three-fold. First, psychology is both science and philosophy. That is to say, it is a science which the student cannot pursue without confronting again and again problems which are commonly regarded as belonging to the field of philosophy. And, just because it is both science and philosophy, it is treated by the universities as neither one nor the other and is allowed to fall between the two stools. In Germany the psychologists have successfully asserted their claim to a fair proportion of the chairs of philosophy. In Great Britain a different course has been followed. Psychology, under the name of mental philosophy, has long been taught in the Scottish and English universities, especially in the former, where such instruction was regarded as a proper part of the education of all students. And until recent years chairs of mental philosophy were filled by men such as Bain, James Sully, James Ward, and Professor G. F. Stout, men who were distinguished primarily as psychologists. But, with the rapid development of psychology and its methods in the last forty years, it has become necessary for him who aspires to all-round competence as a psychologist to study physiology, medicine, ethnology and the laboratory methods. In so doing he marks himself as a man of science; and when, so equipped, he seeks election to a chair of mental philosophy he is met with the dictum that a philosopher rather than a psychologist is needed. The philosophers, warned by the course of events in Germany, have closed their ranks; and though they have allowed a single chair of mental philosophy, namely the Grote Chair of the University of London, to fall to a psychologist, they are not likely to repeat the slip; for even philosophers are human. Thus ignorance— ignorance of all the light that a multitide of workers in many fields of biology have thrown upon the human organism and its functions—has become a necessary qualification of a professor of mental philosophy.

Secondly, there is a certain truth in the oft-repeated dictum that the Englishman dislikes psychology. But it would be absurd to pretend that this is true of all men of English, and still more of all of Scottish, birth; or that it is a native peculiarity of our island. It is true rather that many sons of Great Britain have distinguished themselves by their contributions to psychology. John Locke may justly be called the father of modern psychology. Bishop Berkeley, David Hume, Thomas Reid, David Hartley, the Mills (father and son) Charles Darwin, Herbert

Spencer, Francis Galton, are but a few of those whose fame is largely due to their contributions to this neglected science. English and Scottish psychologists have, prior to the quite modern period of highly organized university research, been the most influential pioneers in the field. In fact, it may be maintained without risk of refutation that psychology is peculiarly congenial to the British genius. All our greatest literature shows clearly the psychological bent of our introverted brooding race. The poetry of reflection upon human nature, and the novel, the study in prose of human character in action, are the most distinctive achievements of English letters. Yet, though the Scots' love of 'metaphysics' (which generally means psychology) is proverbial, it remains true that the orthodoxly educated Englishman does commonly dislike psychology.

There is here a paradox which can be resolved only by taking account of the influence of our orthodox English education. That education for some centuries has consisted, and still largely consists, in the study of Latin, Greek, and religion. Far be it from me to deny that the study of the Latin and Greek languages is an excellent discipline. I am inclined to accept the old claim that it is good, just because it is hard and repugnant to the natural boy. That raises one of the great educational controversies which it is the business of psychology to determine. But, whatever may be the true answer to that question, there can be no doubt that these three studies are based upon authority and severely discourage any tendency to inquire into causal or logical relations, to ask: Why and How? In a mind trained exclusively on these studies any such natural tendency must wither away at an early age. For on every occasion of its exercise it must receive a rude check; to every question which it prompts, the answer comes with an authority against which there can be no appeal—'It is so, and there's an end of it.' And, if the spirit of inquiry would go further and seek boldly to formulate for itself some inductive generalizations that might be applied as explanatory principles, it is again inevitably frustrated and baffled. For in these disciplines there are no laws to be discovered, but only arbitrary rules or conventions, the exceptions to which are more striking than the conforming instances.

Thus that function of the mind which in science plays the predominant role, namely, the explanation of particular instances and events by reference to general principles inductively established, receives no encouragement from the traditional education of the English boy, but rather is nipped in the bud from the moment when it begins to put forth its first tender efforts. It results that a mind so trained, so cabined, cribbed, confined, through all the years of its most rapid development, manifests no tendency to seek for causes and explanations, and no power to form or to grasp such generalizations as science formulates. It was not without good reasons that the Church imposed upon all the studious youth of Europe the minute study of the Latin language; for the Church has always been a sound practical psychologist. No other measure could have conduced so effectively to maintain her authority, to secure in all her children the unquestioning acceptance of her dogmas. And in our British universities, philosophy is still the handmaiden of theology.

It may be objected that, in offering this explanation of the educated Englishman's dislike of psychology, I am explaining too much; for my explanation would impute to him a dislike of all natural science. But the objection is not valid. The conventionally educated Englishman does dislike all natural science. Contemptuous reference to natural science as 'stinks' has long been the fashion at Oxford, and *Scientia delenda est* is the secret maxim of the inner circle. Psychology in this respect has but suffered with the other sciences; but the forces of orthodoxy have borne more heavily against psychology than against the other sciences. When those forces had become reconciled to the Copernican system, the physical sciences appeared to them harmless. The biological sciences, so long as they confined themselves to the collection and classification of species, gave them no offence. But every inquiry into the nature of man was anathema. Anatomy and physiology slowly gained the right to live, because they were so obviously ancillary to the medical and surgical arts. The sciences that deal with the activities of man in the social state and the products of these activities, politics, economics, jurisprudence, philology, were one by one accepted as inevitable; but always with the proviso that they made no inquiry into the nature of man himself. There is even on foot a proposal to institute a chair of politics at Cambridge. But it is safe to prophesy that the occupant of that chair will be as ignorant as he will be scornful of psychology. For the nature of man is a sacred mystery not to be touched by the profane finger of science.

Thus stands the present state of affairs in our universities. The various sciences of man in the social state are recognized and studied, though always with tacit ignoring of their true and proper foundation, the science of human nature itself. But the sciences that would study man himself as a natural denizen of this planet, to wit, anthropology and its departments, ethnology and psychology, remain outside the pale, languishing in byways, unrepresented by chairs in the universities and officially frowned upon. This neglect is the more striking when we reflect that Great Britain has made herself responsible for the welfare of a multitide of races and peoples, and sends her pro-consuls and administrators into all parts of the earth, where, but for the continuing influence of this traditional attitude, they would be able to turn to the advantage of both science and the art of administration the unrivalled opportunities that lie about them on every hand. But these men, with a few shining exceptions, have learnt to regard ethnology and psychology as peculiarly obnoxious varieties of 'stinks'. My experience as a teacher at Oxford during more than a decade showed me only too clearly the truth of the proposition I am here advancing. During those years I endeavoured to induce the honours students to interest themselves in psychology; and I failed conspicuously. It was easy to gain the interest of a number of pass men, men who, like Charles Darwin, had remained recalcitrant to the influences of the traditional system. And those of my students who have continued to work at psychology, and are now making their contributions to the science, are all persons who had not been ground down in the regular mill or who, for one reason or another, had rebelled against it.

The third ground for the present neglect of psychology in Great Britain is something of a reproach to psychology itself. The psychological inquiries of Locke, Berkeley, Hume and other English and Scottish writers of the seventeenth century were actively continued in the following period. In Scotland especially a succession of distinguished men made considerable progress, notably Reid, Adam Smith, Hutcheson, and Dugald Stewart, and were in course of developing a useful and sound psychology. In England, unfortunately, the line of progress was turned aside from the true path by the predominance of two false principles, the principle of hedonism and the principle of the all-sufficiency of the law of association of ideas. Jeremy Bentham uncritically accepted these two principles and dogmatically incorporated them in the utilitarian social philosophy which he taught with so much persuasiveness. The Mills, father and son, associated themselves with Bentham. Later, Herbert Spencer and George Henry Lewes lent their powerful aid; and Alexander Bain allowed himself to be led astray from the sound tradition of the Scottish school and to give adhesion to the very misleading and inadequate psychology of the English school. Hence, when Darwin initiated the modern great development of the biological sciences and laid the foundation of an evolutionary psychology, it was an easy matter for the philosophers to make head against the threatening rise of psychology by pointing to the gross errors and inadequacies of that psychology of association and hedonism which had already become established in England and which was as untenable philosophically as it was sterile and even detrimental for the social sciences. Of course, the true indication was a renewed and more intensive study of psychology leading to a revision of its principles in the light of advancing biological knowledge. But any rope will serve to hang the dog with a bad name; and, under the influences briefly indicated on a foregoing page, the enemies of psychology found no difficulty in pouring scorn upon it and in making its errors seem sufficient ground for casting it out into the wilderness beyond their academic groves.

Medical science in Great Britain illustrates in the most vivid manner my general thesis. As remarked above, medicine and surgery were professions which imperatively demanded some knowledge of anatomy and physiology; accordingly chairs were founded in the universities, and these sciences flourished. But, under the prevailing influences, medical students were given no inducement to consider the mental aspect of the human organism. The average student passed through the whole period of his medical studies without being reminded that man has a mind as well as a body. The consequence is serious enough for both general medicine and surgery, in which the neglect of the mental factor in disorder of the organism long has been and still is a crying evil.

But it is in the field of what we ironically call psychiatry, or psychological medicine, that the traditional British attitude has produced the gravest and the most ludicrous consequences. Our healers of the soul are for the most part excellent administrators of the large institutions in which they work. They learn to make diagnoses according to the current classification of disorders, and to make prognoses that are of considerable value from the point of view of the patient's relatives and friends. They look after his bodily hygiene. But of any attempt to understand the mental factors that have conduced to the state of disorder, the conflicts, the repressions, the disappointments and rebuffs, the thwarted impulses, the faults of character, the failure of adjustment to social situations, they are innocent. And, having no inkling of the role played by these psychogenic factors, they are no less innocent of any attempt to straighten out the tangles in the patient's mind. As regards the all-important early stages of mental disorder when effective help is possible, the state of affairs is even more deplorable. The patient is alone with his secret terrors, his irrational impulses, his obscure distresses; or, if he seeks medical advice, he gets nothing better than a pill or a plaster, the prescription of an expensive sanatorium or a trip round the world.

Those 'psychiatrists' who are disposed to research devote themselves to staining and cutting innumerable sections of the brain after the patient is dead; or, in laboratories equipped at great expense, they make minute studies of the composition of his urine or of the salts contained in his blood. A vast amount of patient labour has been given to such studies. And the total result of them up to the present date, as regards the prevention and cure of mental disorder, may fairly be described in one word—*nil*.

The situation was so well illustrated by my lamented friend, Sir Frederick Mott, one of the most distinguished leaders in the field of psychological medicine, that, at the risk of seeming to belittle the person and work of a man whom I loved and greatly esteemed, I will cite his case. Mott was a devoted admirer and follower of Henry Maudsley, the vigour of whose writing had given his well-known books a foremost place in the English literature of psychological medicine. Maudsley was a materialist of the purest water; and Mott, under the influence of Maudsley and of the traditional English ignorance of the mental aspect of man, went the same way. In spite of occasional gleams of inspiration and common sense, as when he wrote and lectured on the influence of music in improving the state of the mentally afflicted, his outlook remained essentially materialistic. He never grasped the notion of functional disorder, that notion upon which all psychological understanding and treatment of mental disorder essentially depends.

Many a time have I tried in vain to gain for this conception some slight foothold in Mott's mind. When the Maudsley Hospital for the study and active treatment of mental disorders was instituted at Denmark Hill shortly before the War, Mott naturally became the dominant influence in that institution and determined its work along the old and relatively sterile materialistic lines. A great opportunity was thus lost, and British psychological medicine continues to be conspicuously lacking in psychology. Nor is any hope of a remedy in sight. The leading and official positions in the field continue to be occupied by worthy physicians who, victims like their patients of an absurd system, are for the most part innocent of all psychological training and insight. The small sect of faithful Freudians dogmatically preach and practise the teaching of their prophet, rendering thereby the bulk of the profession more resolutely hostile than ever to the psychological point of view; while a very small handful of men in London and the provinces are endeavouring, without official support and in the face of much hostility, to give effect in their private practices to the teachings of psychology.[1]

Before bringing this article to an end, I must refer to one other field in which the neglect of psychology is only less deplorable than in medicine, namely, law. The appearance in the courts of so-called experts in mental diseases who testify, according to the side upon which they are engaged, to the sanity or insanity of the parties concerned,

---

[1] The one bright spot in this scene of darkness is the Institute of Medical Psychology, the result, like the National Institute of Industrial Psychology, of the self-sacrificing efforts of a few enthusiasts.

the accused, the testator, or the claimant, has long been a scandal; and the disputes between lawyers and physicians in this field have been a matter for tears and laughter. The State of Massachusetts has recently given a lead to the whole world by enacting a law which should go far to abate the scandal. Under this new law the court may refer before trial any person accused of criminal action to a panel of expert psychologists for a report upon his mental condition. The court may then have before it on proceeding to try the case, instead of the diametrically opposed statements of two parties of experts each engaged to make out a case for or against the accused, the joint report of the panel of experts paid and engaged by the State to make an unbiased examination and statement. Surely a harmless and perhaps in many cases a useful addition to the resources of the court! But it will be many years before we may hope for any similar addition to the resources of British justice. At the last meeting of the British Association, Dr. James Drever, in his presidential address to the psychological section, stated very temperately the case for some attempt to throw light on the mental condition of persons accused of crime, especially in recurrent cases. The address provoked from the side of the lawyers fervid letters to *The Times*, repudiating the modest suggestion of Dr. Drever and pouring upon psychology and psychologists scorn, contempt, and contumely in the most approved style of the British tradition.

As comment upon this state of affairs I will refer to two passages in a recent issue of the *Morning Post*. One alleges that in Paris hundreds of cases of theft by kleptomaniacs occur ever day. And kleptomania is a curable mental disorder. The other describes how a young man, carrying a dagger and a letter addressed to the Prime Minister of Japan, was arrested by the police. The letter urged the Premier to resign his office; and the young man informed the police that it was his intention 'to hand the letter to the Prime Minister personally and then to stab him if he showed the least sign of turning down the friendly advice'. Surely a case for psychology! But our lawyers, stoutly and virtuously resisting all such fanciful innovations, would turn loose the young man (for he had committed no crime) regardless of the risk to the life of the Prime Minister.

In our disorganized modern societies the number of potential criminals of this type is very large and constantly increasing. I remind the reader of a recent assault upon the Italian Premier by an Irish lady, of the death of Lincoln and its lamentable consequences, and of various other attempts, successful and unsuccessful, by persons too sane to be certified, too insane to be left to their own impulses. And our law and lawyers are perfectly helpless in face of the menace of such persons until after their criminal impulses have attained their goals, until the innocent and generally distinguished victim has received a bullet or a stab. Then, in face of a family bereaved or a nation in tears, our lawyers make ready to hear the interminable disputes of so-called experts as to whether the accused did or did not know the nature of his action. Princes, presidents, and premiers may sigh in vain for protection. Let them be content to know that the law in its majesty will avenge them—if they are so fortunate as to fall victims to assassins whose hired experts cannot persuade the courts of their irresponsibility. In this connexion I cannot forbear to mention the work of Chief Justice Olson, of the Chicago Criminal Court. This enlightened lawyer, firmly persuaded of the close correlation between crime and mental disorder, has secured the able assistance of an expert medical psychologist, by whom all accused persons are examined before judgment is passed. With this assistance and in the light of the startling results obtained, he has been able to convert all or most of the large staff of judges who hear cases in his court to enthusiastic acceptance of psychological examinations of the accused.

I have pointed to grave evils that result from our neglect

of psychology. Is there any remedy? Or must we leave things to the course of nature, hoping that something will turn up to put things right before our national life sinks into hopeless decreptitude, and that somehow we shall muddle through? The evil is the result of neglect extending over generations; and, in the most favourable case, the remedy will require more than a generation for its operation.

As I see it, the main remedial step must be to institute in each of our leading universities, not merely a chair of psychology, but a strong department of psychology, staffed, like any other of the principle scientific departments, by a number of experts, each specializing in one part of the highly diversified field. No one now expects one man to be capable of teaching and research in all the branches of chemistry or of physiology; and the branches of psychology are at least as diversified and require as much special study and training as the branches of any other science. But the men required to staff such departments do not exist; their training and preparation will require many years. And, when they shall have begun their work, many more years must elapse before the young men whom they must train will be able to introduce the psychological point of view into their various fields—industry, economics, politics, education, medicine, etc.—and, by diffusing throughout the community an active interest in the psychological understanding of individual and social problems, will be able to mitigate appreciably some of the evils I have indicated.

The thesis put forward in this article may be briefly stated as follows: The complexity of modern social life has become so great that we can no longer live prosperously by the light of instinct and tradition alone. This complexity has arisen through the discoveries of the physical and biological sciences which have made possible the aggregation of vast numbers of mankind in small areas, have brought all races and nations into intimate contact, have mingled them in heterogeneous masses, and have put at their disposal immense reserves of physical energy and, especially, the means of mutual destruction on a vast scale. While our knowledge of, and control over, physical forces has made immense and ever-accelerating progress, our knowledge of human nature and consequently our power of self-direction and our power of controlling and guiding the development of others has remained rudimentary; because the most difficult and important of the sciences, the science of human nature, has not been assiduously cultivated. In other words, while we have vastly increased our knowledge of the physical world and of the physical basis of life, we have not added appreciably to our wisdom.

This divergence between, on the one hand, physical science and its consequences in the way of increased complexity of social life and, on the other hand, our wisdom or understanding of human nature, is more extreme in Great Britain than in any other country; because, while we have cultivated the physical sciences with great success, we have neglected more grossly than other nations the study of human nature and obstinately continue to ignore what understanding has already been achieved. Hence, more than any other nation, we are threatened by the evils characteristic of our Western civilization, unrest and strife in the industrial world, unhealthy aggregation of population in cities, decay of agriculture, increasing dependence of great numbers upon direct sustenance by the State, an educational system ill-adapted to our needs, a vast increase of neurotic and mental disorders, and a general falling off of the quality of our population. We may hope to remedy these evils only by attaining a greater wisdom than we at present command, a better understanding of human nature, of its weaknesses, its sources of strength, its potentialities, and the conditions favourable to their best development.

XIV
ETHICS OF NATIONALISM[1]

CERTAIN ethical and political problems urgently con-
front the modern world. They are problems which
will have to be met by political action on the widest scale
in the near future—political action which, if it is to be
carried through successfully and confidently for the settlement of the problems I speak of, must conform to principles
recognized as right or ethical. Yet they are problems in
the face of which the ethical principles commonly accepted
by civilized mankind give us no sure guidance.

The ethical principles of all civilizations have had
much in common, in spite of differences in detail and of
emphasis.

To speak the truth, to be mutually helpful and loyal,
to be compassionate, to do no violence to the persons or
property of our neighbours, to practise moderation and
self-discipline—these are the common stock of ethical
precepts, without the cultivation of which, as a strong and
effective moral tradition, no civilization can rise above a
very crude level. No doubt the various civilizations have
emphasized differently these main precepts; each may have
insisted upon certain detailed applications in a manner
peculiar to itself; and such special features of its moral
code may have profoundly affected the course and destiny of
each civilization. Yet, in the main, the differences, as regards
personal conduct of man to man, have been differences of
the moral sanctions rather than differences of precept.

[1] This essay states the problem discussed in my volume *Ethics
and Some Modern World Problems*. The rivalry of ethical systems
here sketched runs parallel in the history of thought with the
rivalry of the two theories of man discussed in the third essay of
this volume, and is indeed another aspect of the same story.

A further common feature of all the historic moral codes is that they have been codes regulating the conduct of individuals in their intercourse with one another, and have had little or nothing to say concerning the relations of group to group, the intercourse of tribe with tribe, of nation with nation.

If we turn from the codes of practical ethics by which men have lived, and by which civilizations and nations have risen and fallen, to the reasonings and speculations of the moral philosophers, we find a corresponding state of affairs. In the main, the moral philosophers have been concerned to define more exactly the true ethical end, the nature of that good which is assumed to be the final goal of ethical endeavour, to refine the current precepts and practices which are the means towards that goal, and to discover the rational sanctions for such precepts and practices. They also, with few exceptions, have been content to discuss the relations of man to man and of the individual to the society into which he is born a member; neglecting those larger ethical problems which arise as soon as one well-defined human group comes into active relations with another.

In short, ethics, both practical and theoretical, popular and philosophical, has been in the main the ethics of the individual.

It is true that some of the ethical systems of the past have given prominence to the relations of the individual to his group considered as a whole, as a living entity with a life, a history, and a destiny of its own, an organism that is more than the sum of the individuals who compose it at each moment of time. And here we find the one important feature that differentiates all ethical systems into two great classes. In this respect, I say, we may properly divide all ethical codes and systems, both popular and philosophical, into two classes, the class of Universal Ethics and the class of National Ethics, ethics of the group, of the tribe, nation, or State. To the former class belong the ethics of Christianity and of Buddhism, and less strictly the ethics of Mohammedanism. Each of these codes is bound up with a religion that aspires to universal dominion; each therefore claims that its rules of conduct are valid for and binding upon all men, and seeks to bring all mankind under the sway of such rules.

On the other hand, the ethical systems of Judaism, of Japan, of China, of Brahmanism, have been national systems; the outlook of each of these systems has been confined to a particular race or nation. And their aim has been, not only to control the conduct of men in relation to one another and for the sake of the welfare and happiness of individuals, but also to regulate the lives of men in relation to the nation or the State; their prescriptions aiming at the welfare of individuals have been modified and complicated by others designed to promote the welfare and the stability of the national group.

This difference may be described by saying that the systems of the one class are universal and individualist, while those of the other are national and political. The difference, the contrast, is illustrated vividly when we compare the ethics of Christianity with the ethics of Judaism.

The Jewish State was a theocracy, and the Jewish people worshipped a national God; their ethical precepts aimed not only to regulate the conduct of men to one another, in. their private relations as individuals, but also and especially to secure the prosperity and the perpetuation of the chosen people, as a national group distinct from all others. The ethical principles of Judaism were ethico-political. On the other hand, the non-political character of the ethics of Christianity was prescribed by its Founder in the command, 'Render unto Caesar the things that are Caesar's'. And, though the various Christian Churches have in later times become affiliated with various States, and though their ethical teaching has been in consequence complicated in some degree by political considerations, the non-political character of the earlier and purer form of Christianity was so well marked as to provoke the resentment of the Roman State.

In this respect the ethics of Greece and of Rome were peculiar. In both cases the popular, the practical, ethical code by which the mass of men lived was essentially ethico-political; for their gods were national gods, and popular ethics and its sanctions were national. The moral philosophers of those States, on the other hand, taught ethical principles and precepts of universal validity; yet they were so far influenced by the spirit of nationality or statehood, by the spirit of national exclusiveness, that they seldom sought to apply their universal principles to the relations of men outside the limits of their own group. Their ethical principles claimed to be generally valid for all men; but the only men generally recognized as men in the full sense were their free fellow-citizens. Their slaves, even those of similar race, as well as the men of other races and nations, remained for the most part outside their purview.

Hence these philosophers failed to achieve any synthesis of ethical and of political principles that could have general validity.

Let us pause here to contemplate the influence of ethical systems of these two opposed types upon the fate of peoples. A national or political ethical system makes for extreme conservatism, for national stability and endurance. It tends to the preservation of the national type, not only by inculcating respect and reverence for the national gods and other national institutions, but also by preserving in some degree the racial purity of the people; for such a system is indifferent to the making of converts, it inspires no missionary enterprises; it is adverse to intermarriage with aliens, and generally adverse also to the admission of aliens to the privileges of citizenship. These effects we see illustrated by the history of China, of Japan, and of Judaism.

China is the supreme example of endurance among

nations; and of that endurance the ethical creed, with its worship of the emperor, its reverence for ancestors, its cult of the family and its hostility to foreign influences, has been, we may feel sure, a main condition—a condition which has preserved the people as a great nation, with all the essentials of its culture, through many centuries, in spite of vast natural calamities of plague and flood and war, and in spite of the lack of natural science and the correlative of that lack, the flourishing of many gross superstitions. Japan repeats the history of China on a smaller scale.

Even more striking illustration of the same influence is afforded by the history of the Jewish people. For there, in the absence of every other condition favourable to national survival, the influence of a strictly national ethical code, backed by strong religious sanctions, has sufficed to preserve the people; and although they were few in numbers, were scattered widely over the face of the earth, and had no national home, it has kept for them something of the character of a nation.

If Greece and Rome failed to maintain their national life for periods comparable to the long endurance of those other peoples, was it not just because the national system of ethics was in each case undermined and fatally weakened by the speculations of philosophers, who taught effectively ethical doctrines incompatible with the rigid conservatism of the old systems? Was it not just for such teaching that Socrates was condemned to drink the hemlock bowl? And, if the old Roman religion and ethics owed their decay less to the speculations of philosophers, was not the downfall of the Roman civilization nevertheless due in the main to other influences of similar tendency? Of these influences, two seem to have been most powerful. First, Rome's success as a conquering power brought her into contact with, and into rulership over, so many peoples of diverse creeds and codes, and ultimately to the absorption of these multitudinous diverse elements within her system, that the old creed and the old ethico-political code, peculiar to and traditional to the small nucleus only of the vast empire, were inevitably swallowed up and their power to guide the conduct of the Roman citizen fatally weakened. Secondly, the spread of Christianity within the empire effected a radical transformation of the ethical system; or, rather, it substituted for the national system one essentially universal and non-political. These two processes of change favoured each the other; and together they destroyed the ethical basis on which ancient Rome had founded and built up her political power. Rome, in short, attempted to assimilate, to Romanize, an immense mass of population of diverse races, creeds, and codes; and in this attempt her ethical system, the source of her power and the foundation of all her greatness, was destroyed.

Systems of national ethics are, by their intrinsic nature, incapable of extension to alien peoples without losing their effectiveness to guide the lives of men. Hence those that have endured have done so only by remaining true to their intrinsic principles, by remaining strictly national and exclusive.

The universal systems and the peoples that have lived under their sway have had a very different history. These systems are by nature assimilative and missionary, seeking to extend themselves over all the world. The three great systems of this type have been so successful that they now include all peoples, save those few which had developed strong national systems before coming into contact and free rivalry with the universal codes. And in the main they have spread by destroying or supplanting the lesser national codes. Since their appearance, each initiated by a single great teacher, the history of the world has been essentially the history of the struggle between these universal systems and the multitude of national systems that had slowly developed during the long ages of the prehistoric period.

The tendencies of a universal ethics are illustrated most

clearly by the history of the Moslem world. The ethical system of Mohammed was planted by him among a people whose tribal creeds and cults were locally restricted and very primitive. It spread with astonishing rapidity, showing a tremendous power of assimilation. Peoples of the most diverse races—white, yellow, and black—and of the most diverse creeds and codes, yielded before its onslaught and were welcomed within the fold; for it accepted all men without question, destroying race-prejudices and national sentiments. It abolished caste and ignored colour, and broke down all barriers that divide man from man; and, what is more important and has been of greater effect in determining the history of the Moslem civilization, it broke down all the barriers that divide man from woman. The Arab mated freely with the Negro and with the yellow races, with the Malay, the Mongol, and the Tartar.

The immensely rapid spread of the Moslem system was due no doubt, in part, to the simplicity of its code and to the relatively simple nature of its sanctions; for these enabled it to appeal effectively to all men. Its code was not too lofty for human attainment; its sanctions were not too remote and ethereal for effective appeal to common human nature. But, most of all, its success was due to the real equality it gave to all its converts. All were made equal in the eyes of God and man, and the career was opened to all the talents. Such multiplicity of contacts of diverse elements of race and culture, such manifold crossings and blendings of human stocks as were thus effected, could not fail to be immensely stimulating to human productivity. And so the rapid spread of the Moslem system was followed by the rise of a civilization astonishing both by the rate of its development and by the richness and variety of its achievements.

In a brief space of time Moslem learning, Moslem science, and Moslem art became predominant on the earth; they covered a broad belt of the old world, from eastern Asia to Spain, with splendid mosques and libraries and universities; while all of Europe that lay beyond their influence still weltered in the chaos left by the breaking down of the Roman civilization.

But this rapid success was followed by a no less rapid decline. The destruction within the Moslem world of the old systems of national ethics rendered possible this rapid flourishing; but it removed at the same time their conserving, stabilizing influences. Soon the brilliance of Moslem civilization was dimmed, a fatal inertia replaced its pristine vigour; and, though its religion still spreads among the more primitive peoples, worsting Christianity wherever they come into direct and fair competition as missionary powers, it has long ceased to add anything of note to the sum of human culture.

It has sometimes been assumed that the Moslem ethos is essentially opposed to progress in the higher things of the spirit. But, in the face of the great and rapid achievements of its early period, we can hardly accept that view. Rather, the history of Moslem civilization implies that a rapid development, soon to be followed by stagnation or actual decay, is its intrinsic tendency. And this two-fold tendency, which its history so clearly displays, is inherent in its ethical system. The universal character of that system and of its religious sanctions, which led it to welcome all comers on equal terms, to override and ignore or destroy all barriers of race and nationality and caste, made for a multitude of stimulating contacts and set free the powers of all its converts from the constricting bands of local and narrow cults and of national or tribal codes.

But the Moslem ethos was lacking in conservative influence. And here we must distinguish widely between influences which are conservative and those which merely clog the wheels of progress and stifle the movements of the mind. Conservation is not the antagonist of progress and of liberalism; it is rather their proper and necessary complement, without which progress and liberalism lead only to early dissolution and decay. The essential expressions of conservatism are respect for the ancestors, pride in their achievement, and reverence for the traditions which they have handed down; all of which means what it is now fashionable to call 'race prejudice' and 'national prejudice', but may more justly be described as preference for, and belief in the merits of, a man's own tribe, race, or nation, with its peculiar customs and institutions—its ethos, in short. If such preferences, rooted in traditional sentiments, are swept away from a people, its component individuals become cosmopolitans; and a cosmopolitan is a man for whom all such preferences have become mere prejudices, a man in whom the traditional sentiments of his forefathers no longer flourish, a man who floats upon the current of life, the sport of his passions, though he may deceive himself with the fiction that he is guided in all things by reason alone.

Such a universal code breaks down also the traditional groupings of mankind; it sets free each man from the control of the group-spirit, which, more than any other influence, renders men loyal members of society, ready to spend and sacrifice themselves for the good of the group, obedient to its laws, and regardful of its future welfare.

In yet a third way, perhaps of greater effect than these other two, the Moslem ethos prepared the stagnation of its own culture. It happened that the Arab people, among whom the Moslem culture took its rise, inhabited a land which lay at the juncture of three continents, the historic homes of the three most distinctive races of mankind—the white, the yellow, and the black—and which was in touch also with the island homes of the Malay race. The breaking down of the barriers of national and racial exclusiveness led to the intermarriage of Moslem converts of all these races. This may have contributed to accelerate the blooming of the Moslem culture, as it certainly contributed to accelerate the spreading of its influence. But the Arab founders freely mixed their blood with that of many other races, and especially with that of the Negro race—a race which never yet has shown itself capable of raising or maintaining itself unaided above a barbaric level of culture. It seems to me probable in the highest degree that this miscegenation, and especially perhaps the large infusion of Negro blood into the peoples bearing the Moslem culture, was a principal factor in bringing about the rapid decline of that civilization.

Now let us consider, from this point of view, the history of the European peoples who became the heirs of the Graeco-Roman civilization. We have noted how the national ethics of Greece and of Rome were sapped and were supplanted by the universal ethics of Christianity. We have now to notice that the peoples of northern and western Europe who came into contact with, and in various degrees under the influence of, the decaying Roman power and the rising power of the Christian ethics, were for the most part organized in strong tribes and rudimentary nations, having their own strong systems of national ethics. And when these contacts took place the Roman civilization was already on the wane; its ethical system was in trans- formation; the national system that had been the foundation of the civil and military power of Rome was already largely destroyed by a system essentially incompatible with, and adverse to, the continuance of that power. This disharmony within the Roman civilization rendered it incapable of dominating the European peoples in the complete way in which Moslem conquerors have dominated their converts. The new converts were only partially converted. They became Christians; but they retained in large measure their national codes and cults. The Englishman became a Christian convert; but he continued to be primarily an Englishman and only secondarily a Christian; and where the dictates of the two systems conflicted those of the national system generally prevailed. The same was true in greater or less degree of all these new bearers of the civilization of Europe. So there grew up the strange anomaly of Christian Europe, a society of nations all of which had accepted the religion of peace and brotherhood, with its universal ethics, yet which were constantly at war with one another.

All the nations of Europe have developed on this twofold ethical basis, have developed ethical codes in which are mixed the incompatible precepts of the universal and of the national ethics. This disharmony of their ethical bases has had profound effects; it has brought certain advantages as well as great disadvantages. Among the advantages we may place first the stimulus to thought and discussion that comes from the conflict of the incompatible elements of the dual code. Where the national ethics holds undisputed sway, as in early Rome, men have no occasion to question its precepts. And where, as in the Buddhist or the Moslem world, a universal code alone rules the conduct of men, there also discussion of ethical principles finds no occasion. But where, as in Athens in its prime, in the later Roman world, and in modern Europe, the two systems are current in imperfect combination, there doubt, questioning, and interminable discussion of ethical principles inevitably occupy men's minds, stimulating them to habits of sceptical inquiry, the effects of which are carried far beyond the bounds of strictly ethical speculation. The progress of European thought and culture has been, no doubt, largely due to this influence.

The imperfect combination of the two ethical systems has been favourable to the progress of European culture in another way. The influence of the universal system has played a great part in bringing about the diffusion of men of European origin over the surface of the globe. Missionaries of Christianity have been, in nearly all cases, very active in the opening of new territories to European colonization. The story of the efforts of the Jesuit missionaries in Canada, Louisiana, South America, Asia, and the Pacific, striking and heroic as it is, reveals only a small part of this vast influence in shaping the present phase of world-history. The colonization of North America was largely due to the conflict between the two systems; for it was this conflict that drove the Pilgrim Fathers to seek new homes across the ocean. They were men in whom the conflict between the two systems became acute and in whom the universal prevailed over the national system. Thus the dual ethics played a great part in bringing about those contacts with strange lands and strange peoples which have reacted so strongly upon the European nations, feeding the appetite for further knowledge, for better means of communication, and for all that was novel, and enriching European civilization with a thousand things and practices brought from the remotest parts of the earth.

In yet a third way the duality of the ethical basis was favourable to progress. While the universal system worked as a liberalizing influence that set men's thoughts and actions free from the bonds which a strictly national system maintains, the persistence beside it of the national systems was a conservative influence which rendered possible the growth of stable nations, each developing its own peculiar variety of institutions and culture, each entering into a stimulating rivalry with the others. Thus was produced that diversity of culture within the bounds of a common civilization which has been a main condition of European progress. If Christian Rome had been strong enough to assimilate completely the tribes and nations within and around the Christian Empire, and had made of Christendom a single great empire based only on Christian ethics, it is probable that its civilization, though it would have bloomed more rapidly, would, like that of the Moslem world, equally rapidly have sunk into apathy and stagnation, if not into actual decay. For, like the Moslem world, it would have lacked the national codes which, while maintaining diversity of cultures, gave strength and stability to the nations as they developed, each acquiring its own peculiar ethos and political structure.

Against these advantages of the dual system we must

set off certain grave drawbacks. The penalty of progress is unrest and a discontent, which, whether we call it divine or merely distressing, contrasts strongly with the peace and whole-heartedness of the saint, whether Buddhist, Moslem, or Christian, and is equally far from the unquestioning devotion of the Samurai warrior, who, in single-hearted acceptance of his national ethics, goes cheerfully and unquestioningly to meet death in the service of his emperor and country.

In the soul of the European two voices have contended for mastery; two claimants for his undivided allegiance have struggled within him, the one proclaiming the duties of the universal Christian code, the other urging obligations of service to his city, his State, his king and country, his nation. And the wars and bloody persecutions which have figured so largely in the history of Europe have been in the main the outcome of the rivalry between the two ethical systems.

With the conversion of the Roman world to Christianity, the two systems of ethics came into open conflict or hardly disguised rivalry in all parts of the empire. The universal system rapidly gained the upper hand. The Church, asserting its claim to be the supreme temporal power, effected a partial synthesis and co-operation of the universal and the national systems of ethics; and, supporting its claim with the tremendous sanctions of the Christian religion, it dominated Europe for more than a thousand years. During this period it suppressed the manifestations of the spirit of nationality, and achieved in large measure a unity of Christendom in which national distinctions seemed in a fair way to disappear. But the spirit of nationality and the old national systems were not dead, though slumbering; and, as the spirit of inquiry began to move again in Europe, men's minds attained to a greater independence, became less subject to the influence of the awful sanctions wielded by the Church. Then the national systems began to assert themselves again, and a tremendous conflict began. The so-called wars of religion were incidents of the resistance offered by the national systems to complete absorption and destruction, of the endeavour to check and throw off the increasing dominance of the Church of Rome.

The spirit of nationality, whose victory in the Reformation ushered in the modern period of European history, has continued to prevail more and more throughout Europe up to the present day; and, more than any other factor, it has shaped the history of the modern world.

The Great War was the culmination of this modern tendency. It was provoked by a nation in which the universal ethics had become completely subordinated to the ethics of nationality, in which the influence of Christian ethics had fallen so low that it failed to restrain and mitigate the boundless aspirations of an unbridled nationalism.

And the Great War has brought no solution of the problem, but rather has accentuated it everywhere. Everywhere, in private conduct and in national policy, we are confronted by the perplexities arising from our dual system of ethics, from the conflict between the claims of nationality and citizenship on the one hand and of the brotherhood of man upon the other.

This unresolved conflict is the essential ground of the present intolerable situation in Europe. France stands out as the embodiment of the spirit of nationality; and most of those who deprecate and condemn her present action are moved in some degree by the spirit of universal ethics. The perplexities of individuals arising from the same source are no less great than the perplexities of nations. The position of the conscientious objector during the Great War was but the clearest illustration of such personal perplexities and dilemmas.

What has been the influence of the speculations of moral philosophers upon the ethical basis of European civilization during the Christian era? With few exceptions, they have thrown their influence on the side of the universal code. This has been true, not only of the great Christian moralists, but also of those who were not specifically Christian. The formula of Kant—'Treat no man as a means, but every man as an end in himself' ; the formula of Bentham and the Utilitarians—'Act for the greatest good of the greatest number' ; the formula of Schopenhauer, which acknowledges acts of loving kindness as the whole sum of moral action—all these are clearly universalist formulas. They take no account of the great fact of nationality; they ignore the obligations and duties that arise therefrom; they are formulas fit for a world that has passed beyond the need for civil government, for national defence, for patriotic self-sacrifice, for loyalty to fellow-citizens or fellow-tribesmen, and to national or tribal institutions. It is true that a few thinkers, notably Machiavelli, Bodin, and Hobbes, have sought to justify and establish the principle of nationality. But they were regarded as political rather than as ethical philosophers; for the world had forgotten the lesson taught by Plato, that the principles of ethics and of politics are inseparable.

The modern world has produced one striking exception to the rule that the moral philosophers have thrown their influence on the side of the universal code, namely, the ethical system of which the philosopher Hegel was the great exponent. Here we have an ethical system propounded by philosophers which threw its whole weight against the universal ethics and on the side of the ethics of nationalism. It was essentially a worship of the State as the highest expression in our world of the universal mind or reason. It taught that the State was that for the sake of which men exist; that each man is before all things a citizen; and that all his ethical obligations derive from his status as a citizen, a member of a larger whole apart from which he is of no value, and has no ethical rights or duties. According to the teachings of this system, a man's conduct is right or moral in so far as he obeys the State, serves it, promotes its welfare, plays a part as a faithful cog in the great machine; but, in so far as his acts may have no relation to the welfare of this larger whole, they are morally indifferent, without ethical significance. The Kantian doctrine is reversed; each man is no longer an end in himself, but solely a means to an end, namely, the welfare of the State. This moral philosophy, being a revival and extreme development of the nationalist system of ethics was eagerly accepted by the Prussian State, in which it took shape; and this State, having elaborated a very efficient system of public instruction, assiduously propagated this code so acceptable to its ambitions; until, after little more than half a century, the national ethics preponderated greatly in influence over the universal system.

We have witnessed and Europe has suffered the terrible effects that may be produced in the modern world by a system of strictly national ethics, unsoftened, unrestrained by an admixture of universal ethics.

This episode of recent history has brought to the front, in public discussion and in private reflection, the great ethical problem that confronts the modern world. It has shown that our civilization can no longer endure upon the dual ethical basis, an ethical hodge-podge of elements mixed from two conflicting and unreconciled systems. The conscience of mankind is profoundly disturbed. Western civilization is sick; its condition is similar to that of the neurotic patient who is torn by conflicting and irreconcilable desires; its moral energies are wastefully consumed by the internal conflict, instead of being devoted to profitable work that would carry our civilization onward to higher levels. The patient suffers from aboulia, or lack of will-power, from various anaesthesias and amnesias, from paralysis, from bad dreams of calamities to come, and from a vague but acute distress. He sees no way of escape, no way in which his conflict may be resolved and his energies once more directed, in harmonious co-operation, towards some clearly envisaged goal. Just as the neurotic patient can be cured only by a complete readjustment of his moral basis, by frankly facing and analysing his problem, by going down to his moral foundations and laying them anew; so also our civilization can be cured, not by any tinkering with symptoms, by moral exhortation, or by sporadic acts of charity to starving peoples, on however great a scale, but only by facing our moral problem, diagnosing its true nature, and thinking out the true solution of it.

The natural unthinking reaction of the earnest Christian or of any man of humane sentiments, in face of the distracted and deeply troubled world, is to denounce the ethics of nationalism as accursed, and to demand that it be wholly swept away to give immediate and undisputed sway to the universal ethics of Christianity. Such a man is apt to assert that, if only all men and all nations would follow strictly the precepts of the Sermon on the Mount, all would be well with the world. Misled by the narrow teachings of the greater number of the moral philosophers, who, ignoring the claims of national ethics, have taught almost exclusively the principles of universal individualist ethics, the greater part of civilized mankind has learnt to regard the universal system as alone ethical or moral, and, while yet practising in various ways the principles of national ethics, never realizes that these also are moral principles that have their valid claim upon our allegiance.

The civilized man of to-day gives a theoretical allegiance to the universal system only; but, when the two systems conflict, he follows in the main the principles of national ethics, justifying such practices, if he at all seeks to justify them, on the ground of urgent practical necessity. And so he repeatedly and constantly finds his practice inconsistent with his professed and consciously accepted ethical principles. And, in the practice of the national ethics under the plea of practical necessity, he lacks the guidance of any mature reflection upon the ethical problems involved. Further, in the advocacy and execution of all national policies, he finds himself hampered, not only by the lack of such deliberately reasoned principles, but also by the fact that such policies are perpetually attacked and opposed by all those who, claiming to speak in the name of morality, urge against such policies the precepts of universal ethics, the only ethics officially recognized and taught as such among us.

Thus the citizen of any one of our modern nations finds himself involved in a situation which is both perplexing and demoralizing. He finds himself supporting national policies which are widely denounced as immoral and which are unmistakably opposed to the generally recognized principles of universal ethics. Yet his good sense forbids him to abandon or to oppose these policies; though he cannot reconcile them with his ethical principles, the only ethical principles that he has been taught to recognize as such.

The earnest Christian who finds himself supporting his nation in war, and perhaps shouldering a rifle in the ranks, illustrates most strikingly this perplexity of the modern mind and this discrepancy between men's practice and their acknowledged ethical principles.

Why, then, cannot we escape from our perplexities by courageously putting into practice the officially and generally accepted principles of universal ethics?

The answer in brief is that the good sense which, in all the foregoing and in many other instances, finds itself opposed to the precepts of universal ethics is not, as so commonly alleged, the expression of mere selfishness and immorality. It is rather the expression of the rival ethics, the system of national ethics, which, though now unformulated and unacknowledged by our moral philosophers, has nevertheless played an essential part in the progress of civilization and still has a very essential part to play in the future; which in fact is required now, as never before in the history of the world, to exert a conservative influence, mitigating and correcting the principles of universal ethics.

The verdicts of common sense or good sense, which are directly opposed to the precepts of universal ethics, need to be philosophically justified and defended against the host of critics who claim to speak in the name of morality. For so long as the champions of good sense are plausibly represented as striving for immoral ends or as using immoral means, their hands are weakened, their resolution is apt to be sicklied o'er with the pale cast of doubt, and their cause is in danger of defeat.

We need, then, in the first place a defence of some of these verdicts of good sense and a demonstration that they are the verdicts of the neglected national ethics; secondly, we need to realize that national ethics is a necessary moral complement of the universal ethics of Christianity; thirdly, we need to harmonize, reconcile, or synthesize the principles of the two systems of ethics which hitherto have remained in open conflict.

Let us go back to the question of the humane man and the sincere Christian, who says, Why not solve our perplexities by boldly and strictly following the precepts of Christian ethics, applying them to the solution, not only of our private personal problems, but also to all public and political problems?

This demand, when it is translated into terms of political action, takes two principal alternative forms. The one form is the ideal of the philosophic anarchists, of Tolstoy, of Kropotkin, and their fellows; the ideal of a world that should need no government, because every man and woman would obey with perfect self-suppression and perfect wisdom the dictates of the universal ethics of human brotherhood. No doubt, if this revolution could be brought about, the state of the world would be improved. But the experience of nearly two thousand years shows that this demand and this hope cannot be fulfilled. They could be fulfilled only if human nature could be radically transformed, in a manner and degree that we know to be impossible. Human nature, the constitution which each of us inherits, the innate endowment of the species *Homo sapiens*, is the product of a long, slow process of evolution; this native basis can be changed only very slowly.

Our innate constitution is not, as John Locke said, and as the optimistic philosophers of the nineteenth century believed, a *tabula rasa*, a clean wax tablet, plastic to receive and to retain whatever form and impress may be given to it. If this doctrine were true, it would follow that we need only to improve the environment in order to transform the whole human race into perfect beings. This was the false philosophy upon which the hopes and the practices of the philanthropists of the nineteenth century were mainly founded.

Human nature, the innate constitution of the species, may more truly be likened to a palimpsest, a tablet that bears the deep and ineradicable impressions of the experience of the race—impressions made during the millions of years in which the race struggled slowly and painfully upward from the intellectual and moral levels of our animal ancestry.

The mass of mankind cannot be made into angels in the course of a few years, nor in the course of a few generations, by any natural process. We must cut our coat according to our cloth; we must seek to develop such an ethical and political system as will effectively harmonize for social ends those energies of human nature that are common to the whole race of man, those ancient instinctive energies that are the very foundation of our being, the springs of all our activities. In short, the ideal of the Christian or philosophic anarchist, of Tolstoy or Kropotkin, the ideal that would do away with all government and all political institutions is, we know, an impossible one. Men need to be governed, need to be members of an organized polity, if they are to realize the best potentialities of their nature. Only by partaking in the life of an organized political community, held together by ancient, firmly rooted traditions, ethical and political, has man risen from savagery; and only by further development and improvement of his ethical traditions and political institutions can he hope to rise above the very modest level he has thus far attained.

A large part of mankind does live under the sway of the universal ethics of Buddhism; and though in some regions, as in China and Tibet, this universal system does not reign alone, but, as with Christian ethics in Europe, is modified by some infusion of national ethics, there are regions in which such modification is but slight; there we may observe the influence of the universal system as exercised in relative purity. Such a region is Burma. And we may fairly turn to Burma to learn what consequences may be expected from such undisputed sway of universal ethics. What, then, is the spectacle presented by the people of Burma? In many respects it is attractive. It has been asserted that the Burmese are the happiest people in the world. They are mild-mannered and gentle, mutually tolerant and forbearing, and singularly free from the more violent vices and crimes, as befits the followers of Buddha. But against this we have to set off their indolence and their intellectual sloth, which have kept the whole people in a condition of stagnation, preventing the development of their civilization beyond a rudimentary level in the spheres of art and literature, and forbidding even the rudiments of scientific culture; so that gross superstition abounds, and the people remains without power to protect itself against the major accidents of nature and the hostility of other peoples. Contemplating such a people we may well be tempted to exclaim with the poet—'Better fifty years of Europe than a cycle of Cathay!'

The second prescription, widely advocated and less remote, perhaps, from the realities of life than the anarchic ideal, is that of a cosmopolitan government, under whose mild rule all national frontiers and national governments should be swept away and mankind should settle down as one happy family to live peacefully for evermore.

By a strange confusion of thought this ideal is often

spoken of and advocated as internationalism. This implies, I say, a strange and puerile confusion of thought, a gross failure to distinguish between two very different systems of ideals, the cosmopolitan and the international ideals. For surely it needs no deep reflection to discern the difference between these two ideals. There can be no internationalism, that is, no settled regime of friendly rivalry and considerate dealing between nations, when all national boundaries shall have been swept away, when nations shall have been abolished and national governments shall have abdicated in favour of one universal parliament of mankind.

This ideal of a cosmopolitan government superseding the functions of national governments and embracing in one great nation all the peoples of the world, is perhaps, unlike the anarchic ideal, not impossible of realization.

But, though it may be a possible system, we have to face the question—Is it a desirable system? Would mankind flourish under any such system, bringing forth the highest and finest fruits of human endeavour?

I have attempted to give a reasoned answer to that question in my *Group Mind*. And I may repeat here very concisely the conclusion to which that investigation led me. The answer is: No—mankind could not continue to flourish and progress under such a cosmopolitan system. In spite of all the drawbacks and dangers inevitably involved in the existence of nations and the flourishing of the spirit of nationality—drawbacks and dangers that are obvious to the meanest intelligence—nations are necessary institutions; for the following reasons:

(1) Man is a social being; he cannot live and thrive alone; and he can be induced to work consistently for the good of his fellow-men, and in harmonious co-operation with them, only by participation in the life of an enduring organized group—a group that has a long history in which he may take pride and an indefinitely long future on which he may fix his larger hopes. Identification of the individual with such a group is the only way in which the mass of

mankind can be brought to live consistently on a plane of altruistic effort and public-spirited endeavour, observing high standards of social conduct such as must be accepted and must prevail in any community, if it is to flourish on a high plane, if it is to maintain and develop a culture worthy in any sense to be called civilization.

(2) Only a group that is completely individualized and self-contained can effectively subdue and turn to the higher uses of social life the egoistic impulses of men in general. Only such a group can find a place and a function for the talents and ambitions of every man who is born into it, making each individual a member of its vital organization; only such a group can give scope and effective stimulus to all the potentialities of each of its members. Any group less than the nation, any such group as a professional or trade association, or a league of socialists or reformers of any kind, even if it be world-wide in its scope, is incapable of doing for its members what the nation can and in various degrees does do for its citizens, in the way of raising their lives above the animal plane of self-seeking or of merely family altruism.

(3) The universal, world-wide, or cosmopolitan State cannot replace the nations in the performance of these elevating functions of nationality, for two good reasons. First, such a cosmopolitan group would be too vast and too heterogeneous to call effectively into play the social potentialities of men in general; men cannot effectively conceive so vast a group, cannot envisage its needs, cannot trace in imagination the effects upon its life of their own efforts and their own sacrifices; they cannot sympathetically share the desires and emotions, the joys and sorrows, of so vast a multitude, most of whom live under conditions which they cannot even remotely imagine, have needs which they cannot understand, and aspirations which they cannot share.

Secondly, even if all this were possible, there would remain a different and equally fatal weakness inherent in the cosmopolitan system. Just as individuals need the stimulus of example, of emulation, and of contact with a variety of types, if their highest powers are to be evoked, so nations and all other groups require similar stimulus; they need the appeal of emulation to evoke their best efforts. And civilization as a whole requires, if it is to progress, the variety of social and political experiment, the varied specializations of collective function and effort, which can be provided only by the rivalry of a number of nations, each developing, under its own peculiar conditions and in accordance with its peculiar racial genius, its own unique, historical process.

In addition to these inherent weaknesses, any cosmopolitan system that might replace the nations, if it were organized upon any principle that could claim to be democratic in any appreciable sense or degree, would suffer a fatal weakness from its mere size. We know now, from the experience of the last century, how great are the difficulties of representative democracy, even when adopted as the working politics of the most stable and experienced nations; how difficult it is to secure any effective voice to minorities; how easily abuses and distortions of the political process arise, and how difficult it is to rectify them when once they have become established. All these difficulties would be magnified immensely under the cosmopolitan system. Such a system could be maintained only as an autocracy; and that, as we know, would offer not the feeblest guarantees, not the faintest prospect, of continued and harmonious development.

The foregoing paragraphs are a highly condensed statement of the argument for nationalism, and for the sentiment of patriotism or national loyalty, as essential conditions of the good life for the masses of mankind. It is fashionable, among those intellectuals who claim for themselves a monopoly of enlightened liberalism and humane sentiment, to decry patriotism as a barbarous survival which, whatever excuse or justification it may have had in the past, can now and in the future work only harm to mankind. This belittling of patriotism is one of the stock features of the repertoire of the cosmopolitan in his attacks upon nationalism. But the more the influence of religion wanes, the more urgently and obviously do we need the influences of enlightened patriotism and of group loyalties of every sort.

If one had to attempt to compare religion with patriotism as influences making for morality throughout the history of mankind, I, for one, should not hesitate to give patriotism the higher place. Fortunately, throughout the development of European civilization, with its dual system of ethics, the dominance of sentiment over logic, so natural to the mass of mankind, has permitted these two great sources of moral effort, religion and patriotism, to co-operate in large measure, in spite of the logical incompatibility of patriotism with the universal ethics.

In order to realize the immensely beneficial influence of patriotism in this *mélange* of religion and patriotism, we have only to turn to the history of a country saturated with religion but devoid of patriotism. Such a country (I speak of the past, not of the present and very recent past) is India. Let us hear what a great critic has to say of this matter. Mr. William Archer, pondering the problems of India's future in the light of its past, writes as follows, in a book which has never been surpassed, I think never equalled, for clarity of vision and humane wisdom on this baffling topic: 'It is not through religion alone that morality can be raised to the temperature at which it passes into our blood and nerve—into the very fibre of our being. All that is needed is to kindle a sentiment . . . of loyalty to something higher than our own personal or family interests —"something, not ourselves, that makes for" or rather demands, "righteousness".'[1] He then writes of 'patriotism as an inspiring principle' as follows: 'Where are we to find in India this "something not ourselves"? To appeal to the Indian masses on the ground of world-citizenship—of their

[1] *India and the Future.*

participation in the onward march of humanity—would be so premature that the suggestion sounds ironic. But may not the necessary stimulus be found in that very idea of India, of the Motherland, which a timorous or merely selfish policy would have us prescribe as seditious? . . . the loyalty of the Indian schoolboy of the near future should be encouraged to attach itself, not merely to his caste or sect, but to his country. Whether we like it or not, this is what will happen—nay, is happening in certain parts of India. It seems to me that the only true wisdom for the Government is to recognize that the inevitable is also the desirable, and to seek in patriotism that reinforcement of character which is falsely declared to be the peculiar property of religion. "Bande Mataram" should no longer be the watchword of sedition, but should be accepted as the inspiring principle of a great effort of national regeneration. It should be the motto, not only of the schoolroom, but of the secretariat.'

These are wise words. India illustrates most forcibly the fact that where nationalism does not exist, or is but feeble, it is necessary to develop it, in order to render a people capable of self-government, to inspire in them the spirit of public service, of devotion to a community wider than the family. The difference between the recent histories of Japan and China has been in the main determined by the fact that the sentiment of patriotism has long been cultivated in Japan much more effectively than in China. In consequence, while the one people seems to be on the road which will lead it to the highest place among the nations, as a leader in civilization and international morality, the other, remaining inert and helpless in all dealings with the outer world and a prey to civil war and to internal disorders of all kinds, is threatened with universal decay.

Nationalism, then, is a great force, the greatest·force in the modern world; and, like other great forces, it is capable of doing much good or much harm, according as it is directed wisely or unwisely. Love of one's country, or patriotism, does not necessarily involve or tend to generate chauvinism, the hatred of other nations; though the two utterly unlike sentiments are often confused through lack of precision of thought and language. It must be admitted also that much of current nationalism is rooted in chauvinism as well as in patriotism. When humanitarians, cosmopolitans, and anarchists denounce nationalism they have in mind, no doubt, that kind of nationalism in which chauvinism plays a prominent part. But their crusade against nationalism is unwise, not only because nationalism (founded in patriotism) is the greatest of forces capable of elevating the masses of mankind, but also because, as all history shows, no such crusade has the faintest prospect of success. In the face of this tremendous and world-wide moral force it is the part of wisdom, not to attempt to oppose or to eradicate it, but to guide it to noble ends, and to purify, with sympathetic understanding, the sentiment of patriotism which should be, and is, its main root and stem.

A system of universal ethics, expressing itself either as a universal anarchy or as a single cosmopolitan world-embracing State, is then not a tenable ideal, not an ideal that can reasonably be made the goal of our endeavour. For, as we have seen, it would, if it were realized under either form, fail to develop or maintain a civilization under which human nature would flourish and put forth its best fruits, realize its potentialities to the full. Under either form, civilization would stagnate; because men would lack conditions essential to the realization of their highest potentialities, both moral and intellectual.

It may be added that not only would either system prove very unsatisfactory, if it could be established, because unsuited to bring out the best that is in human nature, but also human nature is such as to offer immense, perhaps insuperable, difficulties to the perpetuation of any such system. Man is so constituted that he inevitably develops attachments to those of his fellows who are nearest to him, who most resemble him in their customs, their ways of thinking and feeling; with them he finds himself in sympathy and strongly desires to be in sympathy. He prefers their company to that of men less like himself; he is prejudiced in their favour as against all other men; he understands their point of view, because he sympathizes with them. In other words, men in general are incapable of that strict impartiality which the universal ethics requires of them. It is only a rare individual here and there who achieves a truly universal or cosmopolitan attitude; and he generally achieves his impartiality, not by extending his warmer sympathies to all men, but rather by withdrawing from all more intimate relations and becoming equally indifferent to all men, with great loss to his own moral nature and development.

The immense force and wide spread of the spirit of nationality in the modern world illustrate this fundamental trait of human nature. For its rise has coincided with the great improvements in means of communication which have multiplied a thousand-fold the contacts between men of different races and nations. And this multiplication of contacts, instead of destroying or weakening the barriers of nationality, the 'prejudices' of race, the partiality of men for their own kind, has but accentuated these things, fostered their growth, intensified their influences throughout the world; until now these national partialities, these national prejudices and preferences rooted in national sentiments, have become the most powerful political forces of the modern world and, more than any other factors—more even than the immense economic forces of the indus-trial age—have shaped the history of the Western world throughout the last century. The operation of these 'irrational' forces has falsified again and again the economic interpretation of history, and is accountable for the fact that the prophecies of the economist have generally been so wide of the mark. Against these 'irrational' forces the exhortations of the moralists, the lessons of the historians, the prescriptions of the economists, have battled in vain. Human nature has continued to clasp to its bosom its 'Great Illusion' and to be governed by its 'irrational prejudices'. How, then, in face of this leading feature of the history of the modern world, can we rationally hope that a still greater freedom of intercourse and multiplicity of contacts should reverse the tendency to increasing strength of the national spirit? It remains true in general that the more we know of other peoples the more we prefer our own.

The most urgent need of our time is, then, a harmonious synthesis of the two systems of ethics, the universal and the national. For both are indispensable. Such a synthesis is not impossible. As the Ethics of Nationalism, governing the relations of nation to nation, it must be the foundation of Internationalism.

XV
WHITHER AMERICA?[1]

DURING three centuries a vast experiment has been in process in America, the attempt to create a civilization in which freedom, justice and happiness shall be the lot of all who dwell in the land. The attempt has evoked the brightest hopes, not only within America, but far beyond her borders, and has seemed full of promise. America would seem to hold, if we can read the riddle, the answer to the supreme question: Is man so endowed with intelligence and with impulses towards the true, the beautiful and the good that, under the most favourable conditions, he can make life worth living for all men? Or is man still so stupid, so dominated by impulses of greed and lust and fear and anger, so little evolved beyond his animal ancestry in head and heart, that he must continue to fight and scramble for the satisfaction of his primary needs?

We all, then, are deeply concerned with the question raised by the title of this talk, Is the early promise of America in process of fulfilment? Is the civilization of America still following an upward curve? Has it attained a peak or a plateau? Or is the curve, as some critics think, already declining, or even plunging steeply downward?

The most distinctive feature of the life of America at the present time is one which gives it a dramatic, indeed an epic, quality; it is that in America the constructive forces of good and the destructive forces of evil, God and the Devil, are arrayed in open warfare on a scale never before approached in the long story of man's upward

---

[1] A B.B.C. radio-talk in a series, *Whither Mankind?* given in the spring of 1931.

struggle. The battle is joined over an immense front, and both armies are using all the weapons which modern science and great wealth put at their command. To which side does victory incline?

The welfare of a nation requires that certain great goods shall have been attained and shall be steadily maintained. The chief of these are: first, unity and vitality, the strength essential to self-preservation; secondly, freedom and opportunity for all citizens to develop their capacities for action and enjoyment; thirdly, economic security at a reasonably high level; fourthly, the prevalence of order under wise laws efficiently and justly administered; fifthly, the wise use of the advantages accruing from the four great goods just mentioned.

We may best attempt to weigh American civilization in the balance, by considering in how far it has achieved and seems likely to maintain each of these five great goods.

The American nation has achieved in high degree both unity and strength. The many States, separately founded and each claiming sovereignty, have combined and have brought into their great federation, one after another, a number of new States. The Union seems now to be firmly cemented and likely to endure indefinitely. This has been achieved at great cost and only by much effort directed to this end. It has not resulted from some natural and inevitable process of growth; it is, rather, the realization of human aspiration towards an ideal goal. America has thus given to the world a great example of successful combination of sovereign States, of federalization, a process which inevitably involves the yielding up of various rights and powers originally claimed by each. Since it is only by working out some such process of federalization that the many sovereign nations of the world can hope to avoid mutual destruction by the aid of modern physical science and to achieve the benefits of world-wide co-operation, this achievement is of immense value to the world as stimulus, inspiration, and example.

17

The American federal union is now the most powerful political entity the world has ever known, by reason both of its immense economic and financial resources and of its potential military capacity. Like all other great forces, this great unified political force may do great harm or great good to the rest of the world. At present it hangs in the balance, itself the seat of an acute conflict between those who would use it for the advancement of all mankind and those would let the rest of the world go hang, so long as their own country is strong and prosperous.

The possession of this predominant influence in world affairs is so new a thing that even among the well-disposed, the internationally-minded Americans, are many who are timid about using it as they would like to do; they see so many dangers, internal and external, that they shrink from the bold course and do not whole-heartedly work for the ideal end they acknowledge, the federation of all nations. And the attitude towards international affairs most widely prevalent in the mass of the people is one of selfish complacency. They have a childlike belief in the superior virtue of their own nation and in the natural depravity of all others. Conscious of their own security and power, they would hold themselves apart from all contaminating contacts; at the same time they insist upon the maintenance of large armed forces and are ready to use them ruthlessly in support of their economic interests. They steadily refuse to make in any least degree such sacrifices of absolute sovereignty and independence as are necessary, if law and order are to supersede force in the relations between nations.

Fortunately, the present economic depression is teaching both the timid and the selfish that no one nation, not even America, can flourish alone; that 'patriotism is not enough'; that even in international affairs generosity is in the long run better than selfishness. There is ground for hope that the present lesson may turn the scale and bring America definitely to the policy of brave international co-operation, with great moral gain to her people.

The second great good, freedom and opportunity for all, was attained in very large measure in the early days of America, the days of the open advancing frontier. And, though the lustre and the hope of those days have been dimmed, much may still be claimed under this head. In every part of the vast country are splendidly equipped public schools, leading by wide open paths to a multitude of colleges and universities; and these are supported from public and private sources with a liberality unapproached in any other land. Their material equipment is magnificent, in many cases their buildings and surroundings are beautiful.[1] Their teachers are numerous and highly trained. Research of all kinds is carried on by a multitude of keen students. Dramatic, musical, and athletic activities are carried to a high pitch of excellence by a multitude of keen participants. In all the larger cities are growing up splendid museums of science, galleries of art rich in treasures, and well-administered public libraries. In addition there are numerous highly organized and well-supported institutions for all kinds of good social purposes; together expressing a vast amount of benevolence, of good-will, of organized effort for human betterment.

Under these influences, science, art, and literature are flourishing; beautiful public buildings are rising in all parts of the country; the market for serious books is very large and rapidly increasing; first-rate orchestras are many; and the public appetite for learned lectures seems insatiable.

The third great good, so closely connected with the second, namely, economic security at a reasonably high level for the mass of the people, seemed, until very recently, to have been achieved. By dint of hard work and enterprise and courage the Americans have tamed to their uses a vast wilderness; and they have had their reward in wealth unimagined by any other people. This has been possible

[1] Few Englishmen, I think, are aware that the spacious tree-clad campus of many an American college and university is a place of great charm and beauty.

only in virtue of the wonderfully rich natural resources of the country; but natural resources alone do not account for the result. It remains a great triumph of human effort. It has required, as a necessary feature, the securing of relatively satisfactory conditions to the workers in the great industries, conditions that lead labour to co-operate energetically with capital, realizing that their interests are largely coincident, conditions under which labour renders a good day's work for a good day's wage, refraining from sabotage and all the short-sighted practices that so clog and impair industrial efficiency in some other countries.

The economists tell us that, estimated in real wages, the American working man is seventy per cent better off than his English cousin and considerably further advanced beyond the working classes of all other countries. And this corresponds with the impression of general comfort and well-being that one gets in almost every part of the country.

The last stage of this achievement, one of subordinate importance, was the adoption by industry of standardized large-scale or mass production. This may turn out to have been a mistake; and the whole industrial system may require much modification. But, though this question bulks large in present-day discussions, it is, as I said, of minor importance. The natural riches of America and her industrial equipment assure an economic come-back from the present depression.

But there are serious blots upon this picture of strength, unity, freedom, opportunity, and economic well-being largely achieved. There are now large numbers of the people, especially of the more recent immigrants, whose lot is not much better than that of the toiling masses of the old world. The gates of opportunity are growing narrower; personal liberty is everywhere infringed by a growing multitude of laws and conventions, and also by the rapid supplanting of the independent producer, whether farmer, manufacturer, or trader, by great corporations in which the lot of all but a very few is strict obedience to orders received from above; and in some quarters the ruthless application of the power of capital concentrated in a few hands amounts to intolerable tyranny.

More serious still is the fact that these great goods, freedom, opportunity, and a high economic standard of life, brought to so large a proportion of the people and supplemented by a wealth of uplift agencies, are not producing the effects that might have been and generally were expected. The American people undoubtedly show some of the evil effects of luxurious living—a softening of fibre, a too exclusive and increasing concern with trivial pursuits and pleasures, with bridge and golf and cosmetics and cigarettes, with silk stockings and jazz and petting, with costly but trashy movies, with trivial stories and banal shows of many kinds. Present-day America raises the grave question—Can human nature support a high degree of material comfort, of luxury, without undergoing serious deterioration?

Turn now to the fourth of the great goods, social order under wise laws justly administered. Here is the weakest aspect of American civilization. American politics and legislation have been increasingly disappointing. In the early days of the Republic, political life seemed full of high promise. It was led by great patriots, great statesmen and orators, and it achieved great successes. But now it lacks such leaders. Its two-party system has become ineffective through lack of clearly defined differences of principle. It is largely dominated by professional politicians interested in the spoils of office, in graft and pull and patronage. Too often it puts power in the hands of greedy ruffians.

In municipal life these defects are more accentuated; abuses multiply, one scandal succeeds another; and the sporadic efforts of the good citizens to root out the evils subside one after another with but small achievement. Democracy fails in its primary task, namely, to put in power men of capacity and public spirit. Consequently the public services are corrupt; the police often brutal, lawless, and rotten with graft; even the judiciary produces its recurrent crops of scandalous abuses. These defects are largely responsible for the terrible frequency of robbery and crimes of violence.

The frequency of such crimes is not the worst feature of the situation. Far more serious is the fact that crime is a highly organized industry which uses its immense profits for the further corruption of the representatives of law and order, in the legislatures, in the courts, in the Press, and in the police forces; thus maintaining a vicious circle.

This also would be a comparatively minor malady, if the great mass of the people reacted healthily and strongly against it. But no such reaction is provoked; for the public is attuned to lawlessness, is contemptuous of the law, of the legislatures, and of the courts, and has little confidence in the police. These grave symptoms have reached such a point that the civilization of America is threatened at the very foundations on which the whole structure rests. The sapping process cannot go much further without bringing about a general collapse.

If one seeks to formulate the essence of the malady whose symptoms we have noted, the formula must run as follows: here is a great nation which has lost its respect for tradition, for the moral tradition created by the travail of mankind through long ages and gradually embodied in customs, laws, and institutions. The people are attempting to live by the light of reason alone, fantastically over-estimating the power of human reason. It is possible that a nation of supermen, equal to the best in head and heart that our race produces in small numbers, might flourish on that basis. But a nation comprising large masses of low-grade humanity cannot hope so to thrive, especially when it becomes rich, powerful, and luxurious.

Many of the best Americans still cling desperately to the old belief that education in the schools and colleges can remedy all their ills. But the education of school and college cannot suffice. Some sterner discipline of the whole people is needed. Perhaps regeneration may come through the discipline of adversity. Or discipline may be imposed by some Napoleonic personality. More probably the task may be undertaken by an oligarchy, reproducing on a continental scale those groups of good citizens who from time to time in America have combined to restore order in some local scene of anarchy.

In any case, let us not forget that there are many American citizens anxiously watching the course of events, acutely aware of all the forces of evil, and pondering the possibilities of remedy, as well as a multitude of earnest workers in the cause of 'uplift'. They deserve our utmost sympathy and good will. May they succeed in leading the masses of their countrymen back to the paths of public-spirited endeavour, where they shall labour together with clean hands and hopeful hearts, making America in reality what in the eyes of many of her patriots she has seemed to be, a beacon for all mankind, a guarantee that man's future may be better than his past, a promise that his noblest hopes may be fulfilled.

PRINTED BY
JARROLD AND SONS LTD.
NORWICH

METHUEN'S
GENERAL LITERATURE

A SELECTION OF
MESSRS. METHUEN'S
PUBLICATIONS

This Catalogue contains only a selection of the more important books published by Messrs. Methuen. A complete catalogue of their publications may be obtained on application.

ABRAHAM (G. D.)
  MODERN MOUNTAINEERING
    Illustrated.   7s. 6d. net.
ARMSTRONG (Anthony) ('A. A.' of Punch)
  WARRIORS AT EASE
  WARRIORS STILL AT EASE
  SELECTED WARRIORS
  PERCIVAL AND I
  PERCIVAL AT PLAY
  APPLE AND PERCIVAL
  ME AND FRANCES
  HOW TO DO IT
  BRITISHER ON BROADWAY
  WHILE YOU WAIT
                  Each 3s. 6d. net.
  LIVESTOCK IN BARRACKS
    Illustrated by E. H. SHEPARD.
                              6s. net.
  EASY WARRIORS
    Illustrated by G. L. STAMPA.
                              5s. net.
  YESTERDAILIES. Illustrated.
                         3s. 6d. net.
BALFOUR (Sir Graham)
  THE LIFE OF ROBERT LOUIS
    STEVENSON   10s. 6d. net.
                  Also, 3s. 6d. net.
BARKER (Ernest)
  NATIONAL CHARACTER
                         10s. 6d. net.
  GREEK POLITICAL THEORY
                              14s. net.
  CHURCH, STATE AND STUDY
                         10s. 6d. net.
BELLOC (Hilaire)
  PARIS   8s. 6d. net.
  THE PYRENEES   8s. 6d. net.

BELLOC (Hilaire)—continued
  MARIE ANTOINETTE   18s. net.
  A HISTORY OF ENGLAND
    In 7 Vols. Vols. I, II, III and IV
                         Each 15s. net.

BIRMINGHAM (George A.)
  A WAYFARER IN HUNGARY
    Illustrated. 8s. 6d. net.
    SPILLIKINS : ESSAYS   3s. 6d. net.
    SHIPS AND SEALING-WAX : ESSAYS
                         3s. 6d. net.

CASTLEROSSE (Viscount)
  VALENTINE'S DAYS
    Illustrated. 10s. 6d. net.

CHALMERS (Patrick R.)
  KENNETH GRAHAME : A MEMOIR
    Illustrated. 10s. 6d. net.

CHARLTON (Moyra)
  PATCH : THE STORY OF A MONGREL
    Illustrated by G. D. ARMOUR.
                              5s. net.
  THE MIDNIGHT STEEPLECHASE
    Illustrated by GILBERT HOLIDAY.
                              5s. net.

CHESTERTON (G. K.)
  COLLECTED POEMS   7s. 6d. net.
  ALL I SURVEY   6s. net.
  THE BALLAD OF THE WHITE HORSE
                         3s. 6d. net.
    Also illustrated by ROBERT
    AUSTIN.   12s. 6d. net.
  ALL IS GRIST
  CHARLES DICKENS
  COME TO THINK OF IT . . .
                         Each 3s. 6d. net

CHESTERTON (G. K.)—*continued*
GENERALLY SPEAKING
ALL THINGS CONSIDERED
TREMENDOUS TRIFLES
FANCIES VERSUS FADS
ALARMS AND DISCURSIONS
A MISCELLANY OF MEN
THE USES OF DIVERSITY
THE OUTLINE OF SANITY
THE FLYING INN
            Each 3s. 6d. *net.*
WINE, WATER AND SONG 1s. 6d. *net.*

CURLE (J. H.)
THE SHADOW-SHOW   6s. *net.*
            Also, 3s. 6d. *net.*
THIS WORLD OF OURS   6s. *net.*
TO-DAY AND TO-MORROW 6s. *net.*
THIS WORLD FIRST   6s. *net.*

DEXTER (Walter)
DAYS IN DICKENSLAND
            Illustrated. 7s. 6d. *net.*

DUGDALE (E. T. S.)
GERMAN DIPLOMATIC DOCUMENTS,
  1871–1914
In 4 vols. Vol. I, 1871–90.
Vol. II, 1891–8. Vol. III, 1898–
1910. Vol. IV, 1911–14.
            Each £1 1s. *net.*

EDWARDES (Tickner)
THE LORE OF THE HONEY-BEE
Illustrated. 7s. 6d. and 3s. 6d. *net.*
BEE-KEEPING FOR ALL
            Illustrated. 3s. 6d. *net.*
THE BEE-MASTER OF WARRILOW
            Illustrated. 7s. 6d. *net.*
BEE-KEEPING DO's AND DON'TS
                  2s. 6d. *net.*
LIFT-LUCK ON SOUTHERN ROADS
                  5s. *net.*

EINSTEIN (Albert)
RELATIVITY : THE SPECIAL AND
GENERAL THEORY   5s. *net.*
SIDELIGHTS ON RELATIVITY
                3s. 6d. *net.*
THE MEANING OF RELATIVITY
                  5s. *net.*
THE BROWNIAN MOVEMENT
                  5s. *net.*

EISLER (Robert)
THE MESSIAH JESUS AND JOHN THE
BAPTIST
            Illustrated. £2 2s. *net.*

EVANS (B. Ifor)
ENGLISH POETRY IN THE LATER
NINETEENTH CENTURY
                10s. 6d. *net.*

EWING (Sir Alfred), Preside it of
the British Association, 1932
AN ENGINEER'S OUTLOOK
                8s. 6d. *net*

FIELD (G. C.)
MORAL THEORY   6s. *net*
PLATO AND HIS CONTEMPORARIES
                12s. 6d. *net*
PREJUDICE AND IMPARTIALITY
                2s. 6d. *net*

FINER (H.)
THE THEORY AND PRACTICE OF
MODERN GOVERNMENT   2 vols
                £2 2s. *net.*
ENGLISH LOCAL GOVERNMENT
                £1 1s. *net.*

FITZGERALD (Edward)
A FITZGERALD MEDLEY. Edited
by CHARLES GANZ.   15s. *net.*

FYLEMAN (Rose)
HAPPY FAMILIES
FAIRIES AND CHIMNEYS
THE FAIRY GREEN
THE FAIRY FLUTE   *Each* 2s. *net*
THE RAINBOW CAT
EIGHT LITTLE PLAYS FOR CHILDREN
FORTY GOOD-NIGHT TALES
FORTY GOOD-MORNING TALES
SEVEN LITTLE PLAYS FOR CHILDREN
TWENTY TEA-TIME TALES
            Each 3s. 6d. *net.*
THE EASTER HARE
            Illustrated. 3s. 6d. *net*
FIFTY-ONE NEW NURSERY RHYMES
Illustrated by DOROTHY BUR-
ROUGHES.   6s. *net.*
THE STRANGE ADVENTURES OF
CAPTAIN MARWHOPPLE
            Illustrated. 3s. 6d. *net.*

GAVIN (C. I.)
LOUIS PHILIPPE, KING OF THE
FRENCH   7s. 6d. *net.*

GIBBON (Edward)
THE DECLINE AND FALL OF THE
ROMAN EMPIRE
With Notes, Appendixes and Maps,
by J. B. BURY. Illustrated. 7 vols.
15s. *net* each volume. Also, unillustrated, 7s. 6d. *net* each volume.

GLOVER (T. R.)
VIRGIL
THE CONFLICT OF RELIGIONS IN
THE EARLY ROMAN EMPIRE
POETS AND PURITANS
            Each 10s. 6d. *net.*
FROM PERICLES TO PHILIP
                12s. 6d. *net.*

GRAHAME (Kenneth)
THE WIND IN THE WILLOWS
       7s. 6d. net and *5s. net.*
Also illustrated by ERNEST H.
SHEPARD. *Cloth,* 7s. 6d. net.
   Green Leather, 12s. 6d. net.
Pocket Edition, unillustrated.
          Cloth, 3s. 6d. net.
     Green Morocco, 7s. 6d. net.
THE KENNETH GRAHAME BOOK
('The Wind in the Willows ',
' Dream Days ' and ' The Golden
Age ' in one volume).
             7s. 6d. net.
See also **Milne (A. A.)**

HALL (H. R.)
THE ANCIENT HISTORY OF THE
NEAR EAST   £1 1s. net.
THE CIVILIZATION OF GREECE IN
THE BRONZE AGE   £1 10s. net.

HEATON (Rose Henniker)
THE PERFECT HOSTESS
Decorated by A. E. TAYLOR.
7s. 6d. net. Gift Edition, £1 1s. net.
THE PERFECT SCHOOLGIRL
            3s. 6d. net.

HEIDEN (Konrad)
A HISTORY OF THE NAZI MOVE-
MENT   10s. 6d. net.

HERBERT (A. P.)
HELEN   2s. 6d. net.
TANTIVY TOWERS and DERBY DAY
in one volume. Illustrated by
Lady VIOLET BARING.   5s. net.
Each, separately, unillustrated
          2s. 6d. net.
HONEYBUBBLE & CO.   3s. 6d. net.
MISLEADING CASES IN THE COMMON
LAW.   5s. net.
MORE MISLEADING CASES   5s. net.
STILL MORE MISLEADING CASES
          5s. net.
THE WHEREFORE AND THE WHY
' TINKER, TAILOR . . . '
Each, Illustrated by GEORGE
MORROW.   2s. 6d. net.
THE SECRET BATTLE   3s. 6d. net.
THE HOUSE BY THE RIVER
          3s. 6d. net.
' NO BOATS ON THE RIVER '
     Illustrated.   5s. net.

HOLDSWORTH (Sir W. S.)
A HISTORY OF ENGLISH LAW
Nine Volumes.   £1 5s. net each.
Index Volume by EDWARD POTTON.
          £1 1s. net.

HUDSON (W. H.)
A SHEPHERD'S LIFE
     Illustrated.   10s. 6d. net.
   Also unillustrated.   3s. 6d. net.

HUTTON (Edward)
CITIES OF SICILY
     Illustrated.   10s. 6d. net.
MILAN AND LOMBARDY
THE CITIES OF ROMAGNA AND THE
MARCHES
SIENA AND SOUTHERN TUSCANY
NAPLES AND SOUTHERN ITALY
     Illustrated.   *Each* 8s. 6d. net.
A WAYFARER IN UNKNOWN TUSCANY
THE CITIES OF SPAIN
THE CITIES OF UMBRIA
COUNTRY WALKS ABOUT FLORENCE
ROME
FLORENCE AND NORTHERN TUSCANY
VENICE AND VENETIA
     Illustrated.   *Each* 7s. 6d. net.

HYAMSON (Albert M.)
PALESTINE OLD AND NEW
     Illustrated.   7s. 6d. net.
A HISTORY OF THE JEWS IN
ENGLAND
     Illustrated.   10s. 6d. net.

INGE (W. R.), D.D., Dean of St. Paul's
CHRISTIAN MYSTICISM. With a New
Preface.   7s. 6d. net.

JOHNS (Rowland)
DOGS YOU'D LIKE TO MEET
LET DOGS DELIGHT
ALL SORTS OF DOGS
LET'S TALK OF DOGS
PUPPIES
LUCKY DOGS
     Each, Illustrated, 3s. 6d. net.
SO YOU LIKE DOGS !
     Illustrated.   2s. 6d. net.
THE ROWLAND JOHNS DOG BOOK
     Illustrated.   5s. net.

' OUR FRIEND THE DOG ' SERIES
Edited by ROWLAND JOHNS.
   THE CAIRN
   THE COCKER SPANIEL
   THE FOX-TERRIER
   THE PEKINGESE
   THE AIREDALE
   THE ALSATIAN
   THE SCOTTISH TERRIER
   THE CHOW-CHOW
   THE IRISH SETTER
   THE DALMATIAN
   THE LABRADOR
   THE SEALYHAM
   THE DACHSHUND *Each* 2s. 6d. net

KIPLING (Rudyard)

BARRACK-ROOM BALLADS
THE SEVEN SEAS
THE FIVE NATIONS
DEPARTMENTAL DITTIES
THE YEARS BETWEEN

Four Editions of these famous volumes of poems are now published, viz. :—

Buckram, 7s. 6d. net.
Cloth, 6s. net. *Leather*, 7s. 6d. net.
Service Edition. Two volumes each book. 3s. net each vol.

A KIPLING ANTHOLOGY—VERSE
Leather, 7s. 6d. net.
Cloth, 6s. net and 3s. 6d. net.
TWENTY POEMS FROM RUDYARD KIPLING 1s. net.
A CHOICE OF SONGS 2s. net.
SELECTED POEMS 1s. net.

LAMB (Charles and Mary)

THE COMPLETE WORKS
Edited by E. V. LUCAS. Six volumes. 6s. net each.
SELECTED LETTERS
Edited by G. T. CLAPTON.
3s. 6d. net.
THE CHARLES LAMB DAY-BOOK
Compiled by E. V. LUCAS. 6s. net.

LANKESTER (Sir Ray)

SCIENCE FROM AN EASY CHAIR
First Series
SCIENCE FROM AN EASY CHAIR
Second Series
DIVERSIONS OF A NATURALIST
GREAT AND SMALL THINGS
Each, Illustrated, 7s. 6d. net.
SECRETS OF EARTH AND SEA
Illustrated. 8s. 6d. net.

LINDRUM (Walter)

BILLIARDS. Illustrated. 2s. 6d. net.

LODGE (Sir Oliver)

MAN AND THE UNIVERSE
7s. 6d. net and 3s. 6d. net.
THE SURVIVAL OF MAN 7s. 6d. net.
RAYMOND 10s. 6d. net.
RAYMOND REVISED 6s. net.
MODERN PROBLEMS 3s. 6d. net.
REASON AND BELIEF 3s. 6d. net.
THE SUBSTANCE OF FAITH 2s. net.
RELATIVITY 1s. net.
CONVICTION OF SURVIVAL 2s. net.

LUCAS (E. V.), C.H.

READING, WRITING AND REMEMBERING 18s. net.
THE LIFE OF CHARLES LAMB
2 Vols. £1 1s. net.
THE COLVINS AND THEIR FRIENDS
£1 1s. net.
VERMEER THE MAGICAL 5s. net.
A WANDERER IN ROME
A WANDERER IN HOLLAND
A WANDERER IN LONDON
LONDON REVISITED (Revised)
A WANDERER IN PARIS
A WANDERER IN FLORENCE
A WANDERER IN VENICE
Each 10s. 6d. net.
A WANDERER AMONG PICTURES
8s. 6d. net.
E. V. LUCAS'S LONDON £1 net.
THE OPEN ROAD 6s. net.
Also, illustrated by CLAUDE A. SHEPPERSON, A.R.W.S.
10s. 6d. net.
Also, India Paper.
Leather, 7s. 6d. net.
THE JOY OF LIFE 6s. net.
Leather Edition, 7s. 6d. net.
Also, India Paper.
Leather, 7s. 6d. net.
THE GENTLEST ART
THE SECOND POST
FIRESIDE AND SUNSHINE
CHARACTER AND COMEDY
GOOD COMPANY
ONE DAY AND ANOTHER
OLD LAMPS FOR NEW
LOITERER'S HARVEST
LUCK OF THE YEAR
EVENTS AND EMBROIDERIES
A FRONDED ISLE
A ROVER I WOULD BE
GIVING AND RECEIVING
HER INFINITE VARIETY
ENCOUNTERS AND DIVERSIONS
TURNING THINGS OVER
TRAVELLER'S LUCK
AT THE SIGN OF THE DOVE
VISIBILITY GOOD
Each 3s. 6d. net.
LEMON VERBENA
SAUNTERER'S REWARDS
Each 6s. net.
FRENCH LEAVES
ENGLISH LEAVES
THE BARBER'S CLOCK
Each 5s. net.

MILNE (A. A.) and FRASER-SIMSON (H.)—*continued*

TEDDY BEAR AND OTHER SONGS FROM 'WHEN WE WERE VERY YOUNG'   7s. 6d. net.

THE KING's BREAKFAST   3s. 6d. net.

SONGS FROM 'NOW WE ARE SIX'   7s. 6d. net.

MORE 'VERY YOUNG' SONGS   7s. 6d. net.

THE HUMS OF POOH   7s. 6d. net.

In each case the words are by A. A. MILNE, the music by H. FRASER-SIMSON, and the decorations by E. H. SHEPARD.

MORTON (H. V.)

A LONDON YEAR   Illustrated, 6s. net.

THE HEART OF LONDON   3s. 6d. net.

Also, with Scissor Cuts by L. HUMMEL.   6s. net.

THE SPELL OF LONDON
THE NIGHTS OF LONDON
BLUE DAYS AT SEA   Each 3s. 6d. net.

IN SEARCH OF ENGLAND
THE CALL OF ENGLAND
IN SEARCH OF SCOTLAND
IN SCOTLAND AGAIN
IN SEARCH OF IRELAND
IN SEARCH OF WALES   Each, illustrated, 7s. 6d. net.

NEWBOLD (Walton)

DEMOCRACY, DEBTS AND DISARMAMENT   8s. 6d. net.

OMAN (Sir Charles)

THINGS I HAVE SEEN   8s. 6d. net.

A HISTORY OF THE ART OF WAR IN THE MIDDLE AGES, A.D. 378-1485. 2 vols. Illustrated.   £1 16s. net.

STUDIES IN THE NAPOLEONIC WARS   8s. 6d. net.

PARTRIDGE (Eric)

WORDS, WORDS, WORDS !   6s. net.

PETRIE (Sir Flinders)

A HISTORY OF EGYPT
In 6 Volumes.

Vol. I. FROM THE 1ST TO THE XVITH DYNASTY   12s. net.

Vol. II. THE XVIITH AND XVIIITH DYNASTIES   9s. net.

Vol. III. XIXTH TO XXXTH DYNASTIES   12s. net.

Vol. IV. EGYPT UNDER THE PTOLEMAIC DYNASTY
By EDWYN BEVAN.   15s. net.

PETRIE (Sir Flinders)—*continued*

Vol. V. EGYPT UNDER ROMAN RULE
By J. G. MILNE.   12s. net.

Vol. VI. EGYPT IN THE MIDDLE AGES
By S. LANE POOLE.   10s. net.

PHILLIPS (Sir Percival)

FAR VISTAS   12s. 6d. net.

RAGLAN (Lord)

JOCASTA's CRIME   6s. net.

THE SCIENCE OF PEACE 3s. 6d. net.

SELLAR (W. C.) and YEATMAN (R. J.)

1066 AND ALL THAT
Illustrated by JOHN REYNOLDS.   5s. net.

AND NOW ALL THIS
Illustrated by JOHN REYNOLDS.   5s. net.

HORSE NONSENSE
Illustrated by JOHN REYNOLDS.   5s. net.

STEVENSON (R. L.)

THE LETTERS Edited by Sir SIDNEY COLVIN. 4 Vols. Each 6s. net.

STOCK (Vaughan)

THE LIFE OF CHRIST   6s. net.

SURTEES (R. S.)

HANDLEY CROSS
MR. SPONGE's SPORTING TOUR
ASK MAMMA
MR. FACEY ROMFORD's HOUNDS
PLAIN OR RINGLETS ?
HILLINGDON HALL   Each, illustrated, 7s. 6d. net.

JORROCKS's JAUNTS AND JOLLITIES
HAWBUCK GRANGE   Each, illustrated, 6s. net.

TAYLOR (A. E.)

PLATO : THE MAN AND HIS WORK   £1 1s. net.

PLATO : TIMÆUS AND CRITIAS   6s. net.

ELEMENTS OF METAPHYSICS   12s. 6d. net.

TILDEN (William T.)

THE ART OF LAWN TENNIS   Revised Edition.

SINGLES AND DOUBLES   Each, illustrated, 6s. net.

THE COMMON SENSE OF LAWN TENNIS
MATCH PLAY AND THE SPIN OF THE BALL   Each, illustrated, 5s. net.

TILESTON (Mary W.)
    DAILY STRENGTH FOR DAILY NEEDS
                3s. 6d. net.
    India Paper.   Leather, 6s. net.

UNDERHILL (Evelyn)
    MYSTICISM.   Revised Edition.
                15s. net.
    THE LIFE OF THE SPIRIT AND THE
    LIFE OF TO-DAY   ↑ 7s. 6d. net.
    MAN AND THE SUPERNATURAL
                7s. 6d. net.
    THE GOLDEN SEQUENCE
    Paper boards, **3s. 6d.**   net ;
    Cloth, 5s. net.
    MIXED PASTURE :   Essays and
    Addresses   5s. net.
    CONCERNING THE INNER LIFE
                2s. net.
    THE HOUSE OF THE SOUL.   2s. net.

VARDON (Harry)
    HOW TO PLAY GOLF
    Illustrated.   5s. net.

WILDE (Oscar)
    LORD ARTHUR SAVILE'S CRIME AND
    THE PORTRAIT OF MR. W. H.
              6s. 6d. net.
    THE DUCHESS OF PADUA
              3s. 6d. net.

WILDE (Oscar)—continued
    POEMS   6s. 6d. net.
    LADY WINDERMERE'S FAN
              6s. 6d. net.
    A WOMAN OF NO IMPORTANCE
              6s. 6d. net.
    AN IDEAL HUSBAND   6s. 6d. net.
    THE IMPORTANCE OF BEING EARNEST
              6s. 6d. net.
    A HOUSE OF POMEGRANATES
              6s. 6d. net.
    INTENTIONS   6s. 6d. net.
    DE PROFUNDIS and PRISON LETTERS
              6s. 6d. net.
    ESSAYS AND LECTURES   6s. 6d. net.
    SALOMÉ, A FLORENTINE TRAGEDY,
    and LA SAINTE COURTISANE
              2s. 6d. net.
    SELECTED PROSE OF OSCAR WILDE
              6s. 6d. net.
    ART AND DECORATION
              6s. 6d. net.
    FOR LOVE OF THE KING
              5s. net.
    VERA, OR THE NIHILISTS
              6s. 6d. net.

WILLIAMSON (G. C.)
    THE BOOK OF FAMILLE ROSE
    Richly illustrated.   £8 8s. net.

METHUEN'S COMPANIONS TO MODERN STUDIES
    SPAIN.   E. ALLISON PEERS.   12s. 6d. net.
    GERMANY.   J. BITHELL.   15s. net.
    ITALY.   E. G. GARDNER.   12s. 6d. net.

METHUEN'S HISTORY OF MEDIEVAL AND MODERN EUROPE
In 8 Vols.   Each 16s. net.

I.	476 to 911.	By J. H. BAXTER.
II.	911 to 1198.	By Z. N. BROOKE.
III.	1198 to 1378.	By C. W. PREVITÉ-ORTON.
IV.	1378 to 1494.	By W. T. WAUGH.
V.	1494 to 1610.	By A. J. GRANT.
VI.	1610 to 1715.	By E. R. ADAIR.
VII.	1715 to 1815.	By W. F REDDAWAY.
VIII.	1815 to 1923.	By Sir J. A. R. MARRIOTT.

Methuen & Co. Ltd., 36 Essex Street, London, W.C.2
1033

T - #0827 - 101024 - C1 - 186/123/14 - PB - 9781138565289 - Gloss Lamination